Advances in Research on Neurodegeneration, I

Definitions, Clinical Features, and Morphology

Advances in Research on Neurodegeneration

Advances in Research on Neurodegeneration, I

Definitions, Clinical Features, and Morphology

Editorial Board

D.B. Calne
R. Horowski
Y. Mizuno
W.H. Poewe
P. Riederer
M.B.H. Youdim

Birkhäuser
Boston · Basel · Berlin

Donald B. Calne
The University of British Columbia
Neurodegenerative Disorders Centre
Vancouver, B.C.
Canada V6T 1W5

Werner Poewe
Neurologische Abteilung
Universitätsklinikum Rudolf Virchow
D-1000 Berlin 65
Germany

R. Horowski
Schering AG Berlin
Clinical Research
D-1000 Berlin 65
Germany

P. Riederer
Clinical Neurochemistry
Department of Psychiatry
University of Würzburg
8700 Würzburg, Germany

Yoshikuni Mizuno
Department of Neurology
Juntendo University School of Medicine
Hongo, Bunkyo, Tokyo 113
Japan

M.B.H. Youdim
Technion
Israel Institute of Technology
Faculty of Medicine
Haifa 31 096, Israel

Printed on acid-free paper

Birkhäuser

ISBN 0-8176-3631-5
ISBN 3-7643-3631-5

Typeset by TEXniques, Inc., Newton, MA.
Printed and bound by Edwards Brothers, Ann Arbor, MI.
Printed in the U.S.A.

9 8 7 6 5 4 3 2 1

Contents

III. Alzheimer's Disease

IV. Amyotrophic Lateral Sclerosis

V. Discussions

Preface

The neurodegenerative disorders, taken together, are emerging as one of the major causes of morbidity and mortality. The prevalence of Alzheimer's disease, Parkinson's disease, and amyotrophic lateral sclerosis is rising in the elderly, and, in addition, the elderly are increasing, both absolutely and relatively. In this setting, it is appropriate to augment our efforts to understand, treat, and prevent neurodegeneration. With this ultimate goal in mind, a small annual meeting has been established to review the neurodegenerative disorders and attempt to explore how progress might be made. The first "International Winter Conference on Neurodegeneration" (IWCN) took place in Seefield, Austria, in February 1992. Funding was generously provided by Schering AG, Berlin, and for the excellent organization we have to thank Ingeborg Runge.

This publication represents the report of the Seefield meeting. It is intended to be a concise summary of the papers presented and the discussions that took place. The topics for the first IWCN were chosen to provide a broad foundation of clinical and basic science. First, the problem of aging is discussed, as is the challenge of deciding where normality ends and disease begins — by no means a trivial issue. There follows a series of discussions on the classification of neurodegenerative disorders with emphasis on the problem of determining the degree of overlap and the extent to which they represent distinct entities. Evidence is taken predominantly from pathology and clinical neurology. Alzheimer's disease, Parkinson's disease, and amyotrophic lateral sclerosis are considered in turn. From this first IWCN, there emerges a perception of how rapidly the field is changing and how an answer to one problem leads to the formulation of many new questions. Some of these are accessible to experimental study. It is obvious that improvement in clinical and morphological diagnosis, and more knowledge on the natural history of neurodegenerative disease is important for the evaluation of new therapies.

It is the hope of the organizers of the first IWCN that the relatively unstructured, informal nature of the meeting with its restricted number of participants allowed an interchange of ideas that facilitated the identification of those new questions that are of the highest priority. By publishing the proceedings of the meeting, the insights that have been gained will be disseminated to a wider group of investigators.

D.B. Calne
R. Horowski
Y. Mizuno
W. Poewe
P. Riederer
M.B.H. Youdim

List of Contributors

Eva Braak, Department of Anatomy, Theodor Stern Kai 7, D-6000 Frankfurt/Main 70 Germany

Heiko Braak, Department of Anatomy, Theodor Stern Kai 7, D-6000 Frankfurt/Main 70, Germany

Donald B. Calne, Neurodegenerative Disorders Centre, University Hospital, 2211 Wesbrook Mall, Vancouver, B.C., Canada V6T 1W5

Michael Cordes, Neurodegenerative Disorders Centre, University Hospital, 2211 Wesbrook Mall, Vancouver, B.C., Canada V6T 2B5

Marinos C. Dalakas, Neuromuscular Diseases Section, Building 10, Room 4N248, NINDS, NIH, Bethesda, Maryland 20892, USA

J. Day, Neurogerontology Division, Andrus Gerontology Center, Department of Biological Sciences, University of Southern California, Los Angeles, California 90089-0191, USA

C. Finch, Neurogerontology Division, Andrus Gerontology Center, Department of Biological Sciences, University of Southern California, Los Angeles, California 90089-0191, USA

Nobutaka Hattori, Department of Neurology, Juntendo University School of Medicine, Hongo, Bunkyo, Tokyo 113, Japan

C. Huang, Neurogerontology Division, Andrus Gerontology Center, Department of Biological Sciences, University of Southern California, Los Angeles, California 90089-0191, USA

Shin-ichirou Ikebe, Department of Neurology, Juntendo University School of Medicine, Hongo, Bunkyo, Tokyo 113, Japan

Kurt A. Jellinger, Ludwig Boltzmann Institute of Clinical Neurobiology, Lainz-Hospital, 1 Wolkersbergenstraße, A-1130 Vienna, Austria

K. Jendroska, Neurologische Abteilung, Universitätsklinikum Rudolf Virchow, Augustenburger Platz 1, D-1000 Berlin 65, Germany

S. Johnson, Neurogerontology Division, Andrus Gerontology Center, Department of Biological Sciences, University of Southern California, Los Angeles, California 90089-0191, USA

K.W. Lange, Clinical Neurochemistry, Department of Psychiatry, University of Würzburg, Füchsleinstraße 15, 8700 Würzburg, Germany

N. Laping, Neurogerontology Division, Andrus Gerontology Center, Department of Biological Sciences, University of Southern California, Los Angeles, California 90089-0191, USA

P.N. Leigh, University Department of Neurology, Institute of Psychiatry and King's College School of Medicine & Dentistry, De Crespigny Park, Denmark Hill, London SE5 8AF, United Kingdom

Peter A. LeWitt, Clinical Neuroscience Program, 14800 West McNichols, Suite 001, Detroit, Michigan 48235, USA

Yoshikuni Mizuno, Department of Neurology, Juntendo University School of Medicine, Hongo, Bunkyo, Tokyo 113, Japan

T. Morgan, Neurogerontology Division, Andrus Gerontology Center, Department of Biological Sciences, University of Southern California, Los Angeles, California 90089-0191, USA

N. Nichols, Neurogerontology Division, Andrus Gerontology Center, Department of Biological Sciences, University of Southern California, Los Angeles, California 90089-0191, USA

C.W. Olanow, Department of Neurology, University of South Florida, Tampa, Florida 33606, USA

G. Pasinetti, Neurogerontology Division, Andrus Gerontology Center, Department of Biological Sciences, University of Southern California, Los Angeles, California 90089-0191, USA

G.M. Paul, Clinical Neurochemistry, Department of Psychiatry, University of Wrzburg, Füchsleinstraße 15, 8700 Würzburg, Germany

Elaine Perry, MRC Neurochemical Pathology Unit, Newcastle General Hospital, Westgate Road, Newcastle upon Tyne NE4 6BE, United Kingdom

Robert Perry, Department of Neuropathology, Newcastle General Hospital, Westgate Road, Newcastle upon Tyne NE4 6BE, United Kingdom

Werner Poewe, Neurologische Abteilung, Universitätsklinikum Rudolf Virchow, Augustenburger Platz 1, D-1000 Berlin 65, Germany

G. Ransmayr, Universitätsklinik für Neurologie, Anichstraße 35, A-6020 Innsbruck, Austria

P. Riederer, Clinical Neurochemistry, Department of Psychiatry, University of Wrzburg, Füchsleinstraße 15, 8700 Würzburg, Germany

I. Rozosvky, Neurogerontology Division, Andrus Gerontology Center, Department of Biological Sciences, University of Southern California, Los Angeles, California 90089-0191, USA

Barry Snow, Neurodegenerative Disorders Centre, University Hospital, 2211 Wesbrook Mall, Vancouver, B.C., Canada V6T 2B5

Fabrizio Stocchi, Prima Clinic Neurologica, Universita "La Sapienza," Viale Dell'Universita 30, 00185 Roma, Italy

Adrian Williams, University Department of Clinical Neurology, Queen Elizabeth Hospital, Edgbaston, Birmingham B15 2TH, United Kingdom

Hiroyo Yoshino, Department of Neurology, Juntendo University School of Medicine, Hongo, Bunkyo, Tokyo 113, Japan

M.B.H. Youdim, TECHNION, Israel Institute of Technology, Faculty of Medicine, Haifa 31 096, Israel

Part I

Normal Aging of the Nervous System

Chapter 1

Definition of Natural Aging

Barry Snow and Michael Cordes

The definition of normal aging is important for the execution of controlled studies and the separation of disease from the aging process. While the concept of normality may seem simple, several issues make the generation of a normal group difficult. Much of this difficulty surrounds the way we define a group. Most descriptions are of the mean or mode—it is easy to identify a classical case of Parkinson's disease or motor neuron disease. The difficulty arises when we try to define the limits of a condition. Likewise, the concept of normal aging becomes difficult when we attempt to define the lower limits of normal.

The aging process is a combination of genetic programming and external insults. These external insults may include such day-to-day factors as exposure to sunlight or food constituents. They may also include the cumulative effect of minor illnesses or the insidious process of an otherwise inapparent disease. A common working definition of normal aging includes the absence of disease. While this definition implies that the genetic component of aging is "normal," this is by no means clear. Genetically programmed premature aging such as the Hutchinson-Gilford syndrome (progeria) is clearly not normal.

As our ability to detect subtle disease improves, the group of individuals who meet this definition of normal aging (absence of disease) becomes ever smaller to the point that they may have no relevance to the wider population of interest.

Many measurable parameters have been shown to decline with age. These include aspects of cognition, hearing, vision, brain weight, and dopaminergic function. The definition of these declines is strongly influenced by whether the study was cross-sectional or longitudinal and how

ADVANCES IN RESEARCH ON NEURODEGENERATION, I
D.B. Calne et al.

strict was the definition of normal.

Most examinations of normal aging have used cross-sectional methods. Cross-sectional studies have several important limitations. Older individuals already have proven themselves survivors. If the study variable is related to survival, then spurious age effects will be shown. This is selective mortality, or the survivor effect (Rowe, 1977). The effect is introduced if the variable has any effect on survival, place of residence, and willingness or ability to participate in a study. If the variable has changed for the whole population, then this too may produce a spurious age effect. An example may be improved education in younger people, resulting in an apparent decline in mental function with age.

Longitudinal studies also have important limitations. They are time-consuming and expensive. The population must be stable and the methods of testing must not change or drift with time. The individuals also may learn to perform the test and produce an apparent improvement in results.

Both study designs are potentially confounded by subject selection introduced by willingness or ability to cooperate with testing.

The definition of normal, that is, the criteria for subject selection, is central to measuring the effects of normal aging. A common approach is to screen the subjects with medical history, physical examinations, blood biochemistry and hematology, urinalysis, chest x rays, and electrocardiograms (Rowe, 1977; Rowe and Kahn, 1987). Subjects may be excluded on the basis of the finding of any disease process or disease confined to the system in question.

The exclusion of disease processes on the basis of clinical and laboratory test similar to above introduces several important potential biases. First, the selection process may exclude too many individuals. The resulting group may represent a subset of successful aged individuals who are overselected. It is difficult to call such a group normal if they compose a minority of the population (Calne and Calne, 1988). Second, the performance of laboratory tests in particular begs the question of how exhaustive such testing should be. Why stop at simple tests? This issue is particularly important when exploring the concept of subclinical neurodegenerative disease. If neurodegenerative diseases do have a preclinical phase, then a group of normal subjects potentially includes some of these patients.

The Effect of Age on Dopaminergic Function

Several lines of evidence point to a decline in the function of the nigrostriatal dopaminergic system. Substantia nigra cell counts decline with age (McGeer et al., 1977; Gibb and Lees, 1991); striatal dopamine also decreases (Riederer and Wuketich, 1976; Carlsson and Winblad, 1976). Certain features of aging also bear some resemblance to Parkinson's disease such as gradual impairment of motor function and neurophysiological indices, including long-latency reflexes (Teräväinen and Calne, 1983). These factors do not necessarily indicate reduced dopaminergic function, however. The nigrostriatal dopaminergic system has a large capacity for compensation including increased activity and amount of tyrosine hydroxylase and nerve terminal sprouting (Zigmond et al., 1984; Wolf et al., 1991).

The best *in vivo* method for measuring dopaminergic function in humans is with positron emission tomography (PET) using the tracer [^{18}F]-fluorodopa (FD) (Garnett et al., 1983). FD is taken up by nigrostriatal dopaminergic neurons, decarboxylated to fluorodopamine and retained, probably in nerve terminal vesicles. Decarboxylation and storage are the primary influences on dopaminergic activity measured with FD PET. This method does not measure the activity of the rate-limiting tyrosine hydroxylase. Because decarboxylation is not part of the regulatory process of dopaminergic activity, FD PET seems to be a relatively stable index of the basal functional integrity of the dopaminergic pathway. It relates at least in part to the number of remaining dopaminergic neurons (Chieuh et al., 1986). FD PET has proved capable of distinguishing Parkinson's disease from normal, and of identifying individuals with subclinical dopaminergic deficits (Calne et al., 1985; Nahmias et al., 1985).

Although post-mortem studies of substantia nigra cell numbers and striatal dopamine suggest that FD PET should show a decline with age, it is conceivable that compensatory mechanisms eliminate this effect. A number of studies have asked the simple question at to whether dopaminergic function as measured by FD PET declines with age.

The first study was performed in Vancouver by Martin et al. (1989). The study was cross-sectional and no screening methods were applied to the individuals other than that they did not carry a diagnosis of any disease, did not take medication, and were able to lie still for the 2- to 3-hr duration of the PET study. This study of 10 patients showed a significant decline over the 20- to 80-year age range. The decline was linear and at a rate similar to the decline in substantia nigra cell numbers reported by McGeer et al. (1977).

A subsequent study using similar methods was reported by Sawle et al. (1990). They studied 26 patients aged between 27 and 76 years. They found no age relationship. Their subjects were selected on the basis of absence of a history of neurologic disease or extrapyramidal signs. Although not one of the selection criteria, all subjects happened to score 30 of 30 in the Mini Mental Status Questionnaire and no decline was seen in the long-latency reflexes. This suggests that these individuals were successful agers.

As part of our PET program in Vancouver, we have continued to collect FD scans on normal subjects. These individuals were recruited as part of a separate study on successful aging. All had normal results on detailed clinical and laboratory examination. With the addition of these individuals to those from Wayne Martin's study, an age-related decline was apparent in the total of 28 subjects.

We have also performed a longitudinal study with FD PET. Seven individuals have been scanned approximately 3 years apart. There was a significant decline at a rate similar to that reported by Martin et al. (Bhatt et al., 1991). Because the initial scans in this series were taken at a time when PET methods were changing rapidly, a simplified method of data analysis had to be used in comparison with the studies of Martin et al. and Sawle et al., 1990; it is therefore possible that the different results may have related to the use of different methods.

In an attempt to resolve the issue of an age effect on FD PET, we have initiated yet another study. The design involves scanning individuals aged between 20 and 30 years old and their grandparent aged between 70 and 80. Such a design with two groups has greater statistical power than scanning all comers. The selection of grandparents and grandchildren was an effort to reduce genetic heterogeneity. Longevity is strongly familial—presumably the grandchildren were less likely to be potential poor survivors who might have initially low dopaminergic function. In addition, we reasoned that the groups have a greater chance of starting out with similar levels of dopaminergic function. To date we have scanned four pairs for a total of eight subjects. We have found a decrease in FD uptake within each pair and an overall statistically significant decline.

How do we reconcile these findings? It is possible that the cross-sectional groups that have shown no age relationship have included a large group of subjects who have aged successfully and have not lost dopaminergic function. In addition, it is possible that younger subjects with lower levels of dopaminergic function as a group have reduced life expectancy. If these points were true, then these studies would be compatible with the significant age relationships shown in the longitudinal and the grandparent–grandchild cross-sectional study.

Another possibility is that the studies that have shown an age relationship somehow included individuals with subclinical dopaminergic deficits. Such deficits have been shown in individuals exposed to MPTP and in Guamanians with ALS but no parkinsonism (Snow et al., 1990; Calne et al., 1985). We also have observed a clinically normal individual with reduced FD uptake who went on to develop parkinsonism (Snow et al., 1991). A group of clinically normal Guamanians scanned as controls for the above study showed a marked decline with age. The older individuals presumably had silent pathology as found at post mortem in older Guamanians (Anderson et al., 1979).

Defining the limits of normal aging, then, depends very much on the criteria used to define the study subjects. It is quite possible to get into a circular trap whereby a particular test is the only method to demonstrate individuals with subclinical disease—yet these individuals need somehow to be excluded to define the range of normal for that particular test. In some cases such a subgroup may be recognized by a skew or double peak in the distribution of results in the older age group, with the second peak representing the subclinical diseased group.

The process of excluding individuals with subclinical disease also has important practical and philosophical implications (Calne and Calne, 1988). It is possible that such a process makes the remaining normal group so different from the general population that they have no relevance to a particular study. This has led to the idea that perhaps different normal groups should be defined for different circumstances (Rowe and Kahn, 1987; Calne and Calne, 1988; Calne et al., 1991).

Subjects for studies may be classified as successful agers, usual agers, diseased, and impaired (Rowe and Kahn, 1987). Successful agers are those who have survived without particular age-related decline; any change is largely genetically based and there has been little damage to the system of interest by exogenous factors. Usual agers include that majority of the population who have a combination of genetically programmed changes plus declining function due to accumulated external insults. Diseased individuals have declining function due to some identifiable process. This may be asymptomatic, such as the normal Guamanians who have neurofibrillary tangles at post mortem. Impaired individuals have symptomatic disease.

Dividing control groups as above may be difficult or impossible. Attempts should be made to do so, however, as each group has application for different studies. To define the range of normal for a diagnostic test, successful agers should be used to define the normal range. In the case of FD PET, this is the range that would be used to identify individuals who might have subclinical dopaminergic deficits. A group of usual agers would make up the control group for descriptive epidemiologic studies. Usual agers also would make up the study population of an intervention for neuroprotection or vascular protection, such as prophylactic aspirin for stroke and heart disease.

It also may be informative to compare these different control groups. In the case of FD PET, if it is true that successful agers do not show an age-related decline in FD uptake while usual agers do, then an analysis of the differences between the groups may provide an insight into the factors that lead to the nigral damage of Parkinson's disease.

References

Anderson FH, Richardson EP, Okazaki, H, Brody JA (1979): Neurofibrillary degeneration on Guam: frequency in Chamorros and non-Chamorros with no known neurological disease. *Brain* 102:65–77.

Bhatt M, Snow BJ, Martin WRW, Pate BD, Ruth TJ, Calne DB (1991): Positron emission tomography suggests that the rate of progression of idiopathic parkinsonism is slow. *Ann Neurol* 29:673–677.

Calne DB, Calne JS (1988): Normality and disease. *Can J Neurol Sci* 15:3–4.

Calne DB, Eisen A, Meneilly G (1991): Normal aging of the nervous system. *Ann Neurol* 30:206–207.

Calne DB, Langston JW, Martin WRW, et al. (1985): Positron emission tomography after MPTP: observations relating to the cause of Parkinson's disease. *Nature* 317:246–248.

Carlsson A, Winblad B (1976): Influence of age and time interval between death and autopsy on dopamine and 3-methoxytyramine levels in human basal ganglia. *J Neural Transm* 38:271–276.

Chieuh CC, Firnau G, Burns RS, et al. (1986): Determination and visualization of damage to striatal dopaminergic terminals in 1-methyl-4-phenyl-1,2,3,6-tetrahydropyridine–induced parkinsonism by [^{18}F]-labeled 6-fluoro-L-dopa and positron emission tomography. *Adv Neurol* 45:167–169.

Garnett ES, Firnau G, Nahmias C (1983): Dopamine visualised in the basal ganglia of living man. *Nature* 305:137–138.

Gibb WRG, Lees AJ (1991): Anatomy, pigmentation, ventral and dorsal subpopulations of the substantia nigra, and differential cell death in Parkinson's disease. *J Neurol Neurosurg Psychiatry* 54:388–396.

Martin WRW, Palmer MR, Patlak CS, Calne DB (1989): Nigrostriatal function in man studied with positron emission tomography. *Ann Neurol* 26:535–542.

McGeer PL, McGeer EG, Suzuki JS (1977): Aging and extrapyramidal function. *Arch Neurol* 34:33–35.

Nahmias C, Garnett ES, Firnau G, Lang AE (1985): Striatal dopamine distribution in Parkinsonian patients during life. *J Neurol Sci* 69:223–230.

Riederer P, Wuketich ST (1976): Time course of nigrostriatal degeneration in Parkinson's disease. *J Neural Transm* 38:277–301.

Rowe JW (1977): Clinical research on aging: strategies and directions. *N Engl J Med* 297:1332–1336.

Rowe JW, Kahn RL (1987): Human aging: usual and successful. *Science* 237:143–149.

Sawle GV, Colebatch JG, Shah A, Brooks DJ, Marsden CD, Frackowiak RSJ (1990): Striatal function in normal aging: implications for Parkinson's disease. *Ann Neurol* 28:799–804.

Snow BJ, Martin, WRW, Calne DB (1991): A case of subclinical idopathic parkinsonism detected with fluorodopa PET. *10th Int Symp Parkinson's Dis* (Abstract).

Snow BJ, Peppard RF, Guttman M, et al. (1990): Positron emission tomographic scanning demonstrates a presynaptic dopaminergic lesion in Lytico-Bodig. *Arch Neurol* 47:870–874.

Teräväinen H, Calne DB (1983): Motor systems in normal aging and Parkinson's disease. In: *The Neurology of Aging*, Katzman R, Terry RD, eds. Philadelphia: F.A. Davis.

Wolf ME, LeWitt PA, Bannon MJ, Dragovic LJ, Kapatos G (1991): Effect of aging on tyrosine hydroxylase protein content and the relative number of dopamine nerve terminals in human caudate. *J Neurochem* 56:1191–1200.

Zigmond MJ, Acheson AL, Stachowiak MK, Stricker EM (1984): Neurochemical compensation after nigrostriatal bundle injury in an animal model of preclinical parkinsonism. *Arch Neurol* 41:856–861.

Chapter 2

Molecular and Neuroendocrine Mechanisms in Age-Related Astrocyte Activities and Neurodegeneration

C. Finch, J. Day, C. Huang,
S. Johnson, N. Laping, T. Morgan,
N. Nichols, G. Pasinetti, and I. Rozosvky

This review addresses findings from this laboratory on the molecular mechanisms of age-related astrocytosis and neurodegeneration in humans and laboratory mammal models. We apologize for neglecting work of other colleagues in this brief article; major reviews are cited that give access to the large literature supporting many of the statements made herein. A major and unresolved question in the biology of aging concerns the factors in age-related astrocyte activities and neurodegeneration that are intrinsic versus extrinsic in origin. First, we suggest a framework that helps conceptualize these phenomena among the many complex manifestations of aging.

In general, a set of aging changes occurs throughout different species and populations of mammals that can be described as *canonical*. Most canonical age changes can be designated as intrinsic, since they occur in virtually all environments (Finch, 1990). *Intrinsic* aging changes include those that occur in the widest range of environments and that therefore can be said to be preprogrammed at some level of the organism by the genome. *Extrinsic* aging changes can be ascribed to specific environmental causes or influences, such as unfavorable diets, toxins, or mechanical trauma.

ADVANCES IN RESEARCH ON NEURODEGENERATION, I
D.B. Calne et al.

Examples of canonical intrinsic aging changes in mammals include midlife reproductive senescence through exhaustion of the ovarian oocyte pool and slowed psychomotor reaction time. Nonetheless, aging changes in any cell need not be intrinsic to that cell, and could reflect factors in the physiological environment. For example, ovarian aging occurs through finite numbers of ovarian oocytes and hormone-producing follicles that are produced during embryogenesis; the exhaustion of the follicular and oocyte pool at menopause is thus a consequence of developmental programming. However, oocyte loss during adult life is slowed by hypophysectomy (Finch, 1990), which first suggested that the loss of nondividing and irreplaceable cells could be influenced by hormones. From these features of ovarian aging, we also learned that many age-related changes in estrogen target cells at menopause are not intrinsic, and can be reversed or prevented by maintaining estrogen levels through hormone supplements when steroid production fails at menopause (vom Saal and Finch, 1988).

Alternatively, factor in the external environment have many influences on aging. These influences are shown by wide individual variations in the extent of vascular disease and its consequences to brain function in humans. Individual differences may represent effects of environmental variables such as diet, drugs, and stress. But individual patterns of brain aging also may involve vulnerabilities from susceptible genotypes, as in the hereditary lipidemias. Besides the lipidemias, other genetic predispositions of neurodegeneration are emerging, for example, dominant genes for Alzheimer's disease (AD) and possibly Parkinson's disease (PD). Whether the age at onset and course of hereditary AD and PD are subject to environmental influences is unknown.

Astrocyte Activities During Aging

Hormonal interaction

At the cell level in the brain, several age-related changes are good candidates for canonical changes. One of the best documented is the increase of astrocytic activity, as measured by the numbers of reactive astrocytes and the content of glial fibrillary acidic protein (GFAP), an astrocyte-specific intermediary filament protein. Increased astrocytic activity begins at least at midlife in some brain regions of laboratory rodents (e.g., the hypothalamus of female mice and rats) (Schipper et al., 1981), where there is histological evidence for increased phagocytic activity, as judged by cell enlargement and dense inclusion bodies. Using GFAP as an index, there is increased astrocytic activity by the average life span in the hippocampus and other brain regions of male mice and rats (Goss et al., 1991). Similar trends are found by histological, neurochemical, and morphometric analyses in mice and rats (Landfield et al., 1978; Finch and Landfield, 1985; O'Callaghan and Miller, 1991) and humans (Hansen et al., 1987). Many more species must be examined to establish that these astrocytic changes are canonical.

Astrocytic hyperactivity is classically identified with neurodegeneration or injury, as observed in the surrounds of degenerating neurons after stroke, AD, or PD. This apparently ubiquitous increase of astrocytic activities could imply widespread age-related death or regression of neurons. However, we hesitate to conclude that astrocyte activation always requires irreversible neuron damage. First, there is considerable controversy about the extent of age-related neuron death in the absence of specific lesions like those of AD or PD. While age-related neuron death may be common in such regions as the toxin- and hypoxia-sensitive hippocampal pyramidal neurons of older rodents and humans, there is no evidence for ubiquitous large-scale neuron death during aging in the absence of specific lesions (Finch and Morgan, 1990).

Many experimental lesions also cause rapid increases of GFAP mRNA in various brain regions: deafferenting lesions (e.g., perforant path transsection) (Poirier et al., 1991; Day et al., 1990; Laping et al., 1991). We also note that lesions in organs remote from the brain can cause elevations of GFAP (e.g., liver from humans with hepatic encephalopathy) (Norenberg, 1987). The elevations of brain GFAP in mice with terminal wasting diseases (Goss et al., 1990) further shows this phenomenon and emphasizes the exquisitely sensitive regulation of GFAP. The variety of factors and conditions that cause changes in astrocyte morphology suggest that the usual terms *hyperactivity* or *astrocytosis* are vague and need refinement.

We are investigating roles of hormones in age-related changes of astrocyte activities. For example, castration of adult male rats causes increases of astrocyte activities, as judged by Northern blot hybridization and immunocytochemical increases of GFAP mRNA and protein in the dentate molecular zone of the hippocampus (Day et al., 1990). There is no evidence that castration of adult rats causes dentate neuron death. However, the presence of sex steroid receptors on entorhinal cortex neurons that project through the perforant path to the dentate molecular zone (Simerly et al., 1990) might cause steroid-dependent synaptic remodeling with involvement of these afferents after castration. These results are intriguing because they suggest that astrocytes can undergo major changes in activity in the absence of neurodegeneration. Furthermore, we may now consider whether such changes in astrocyte activities can influence neuronal activities.

An example of altered astrocytic activities involving ovarian steroids is the retardation of astrocytic hyperactivities in the arcuate nucleus of the hypothalamus at middle age in female rodents by ovariectomy (Schipper et al., 1981); these changes are part of the ovary-dependent neuroendocrine aging syndromes of female rodents, which are retarded by ovariectomy and accelerated by prolonged exposure to sustained but physiological levels of estrogens (Finch et al., 1984; Kohama et al., 1989; Mobbs et al., 1992). The histological evidence in Schipper et al. (1981) suggests increased phagocytic activities. As noted above for hippocampal astrocytes, there is no evidence for associated death of neurons, particularly the GnRH-containing neurons in the hypothalamus (Kohama et al., 1992; Hoffman and Finch, 1986). Furthermore, GFAP mRNA changes regularly during the estrous cycle in the mouse arcuate nucleus, with major decreases after the peak estrogen levels at proestrous (Kohama et al., 1991).

Androgens and estrogens are well known to influence the patterns of dendritic growth during development. These effects may be direct, since estrogens have rapid effects on the outgrowth of cultured embryonic neurons in culture (Johnson and Brinton, 1989). One working hypothesis is that steroid-induced remodeling of viable afferents can provoke astrocytic responses. Thus, neurons in many hormone-sensitive brain regions could change in response to age-related changes in gonadal hormones.

However, there is some evidence that astrocytes can respond directly to sex steroids (e.g., estrogen increases the secretion of NGF in cultured C6 gliomas) (Perez-Polo et al., 1977). In mixed cultures of astrocytes and oligodendroglia from neonatal rats, the biosynthesis of progesterone increased sharply during the differentiation of oligodendrocytes, which are implicated as the main steroid-synthesizing cells (Jung-Testas et al., 1989). Much remains to be investigated about glial capacities for sex steroid biosynthesis and catabolism.

Glucocorticoids also influence hippocampal aging. For example, hippocampal astrocytes show age-related hyperactivity that is retarded by adrenalectomy and accelerated by corticosterone treatment (Landfield et al., 1978; Sapolsky, 1990). Contrary to expectations from these histological studies on astrocytes, hippocampal GFAP mRNA is decreased by corticosterone and increased by adrenalectomy during short-term experiments (Nichols et al., 1990; Laping et al., 1991). The divergence of our studies with mRNA from those using silver stains for astrocytic activity (Landfield et al., 1978) may not be as paradoxical as it seems, since cytoskeletal staining can reflect changes in molecular organization as well as in the absolute levels of proteins. Together, the studies on sex steroids and glucocorticoids suggest neuroendocrine mechanisms in aging changes of astrocytes that need not require neuron death.

Preliminary data from nuclear run-on studies indicate that these effects of corticosterone in decreasing GFAP mRNA are at the transcriptional level (Laping, Johnson, and Finch, unpublished). The 5′-upstream region of the GFAP gene contains many putative response elements that are candidates for integrative effects of diverse physiological regulators, for example, for estrogens (ERE), glucocorticoids (GRE), TGF-β inhibitory element (TIE), nuclear factor 1 (NF-1), and activator protein 1 (AP-1), which mediate stimulation of transcription by TGF-β and cAMP (CRE). Other hormonal and physiological axes could be involved in the regulation of GFAP mRNA. These response elements are recognized by reference to consensus sequences of response elements from other genes. Features of the GFAP promotor studied so far concern initiation sites and tissue-specifying sequences (Nakatani et al., 1990; Miura et al., 1990; Balcarek and Cowan, 1985). Ongoing work shows the functionality of GREs in engineered constructs of 5′-GFAP sequences with a reporter gene (Huang et al., 1993).

We do not know which chemical or electrical signal(s) triggers rapid increases of GFAP mRNA in response to lesions of the perforant path that cause degeneration of afferent terminals in the hippocampus. The rapid elevations of TGFβ-1 mRNA in the hippocampus after perforant path lesions implicate this cytokine in response to neurodegeneration (Nichols et al., 1991), especially since TGF-β1 can stimulate GFAP mRNA and protein when given intraventricularly (Laping et al., 1991; Nichols et al., 1991) or to primary astrocyte cultures (Lindholm et al., 1990; Sakai et al., 1990). Besides TGF-β1, many of the responses to hormones, electrical activity, or neurodegeneration also could be mediated by second messengers, which in turn can activate early intermediate genes such as fos and jun; fos and jun combine to form a heterodimer that binds to AP-1 regulatory sites in the promotor of many genes, potentially including the AP-1 site in GFAP.

The increase of GFAP mRNA and protein after castration of young adult male rats has led us to examine effects of aging. In the F344 male rat, plasma testosterone decreases strikingly after middle age. These decreases are more extreme than generally observed during aging in mammals and may be associated with the unusually high incidence of benign testicular tumors in the inbred F344 rat strain. There is no canonical male aging change equivalent of menopause; spermatogenesis continues in most males throughout the life span (Finch, 1990). Moreover, hippocampal GFAP mRNA increased with age, with strong correlations in individual aging rats between GFAP mRNA and plasma testosterone (Nichols, Day, Laping, and Finch, unpublished). Another candidate is TGFβ-1, which shows age-related increases of mRNA in the hippocampus (Nichols and Finch, unpublished); as noted above, TGFβ-1 stimulates GFAP mRNA, and can do so directly on astrocytes. There is evidence for age-related loss of synapses in the dentate gyrus (Geinisman and Bondareff, 1976), which is consistent with electrical indications for loss of perforant path inputs (Foster et al., 1991). Therefore, we also may consider the possibility that TGFβ-1 elevations during aging are consequent to perforant path remodeling, in which changes in testosterone could be secondary.

The age-related increase of plasma corticosterone that is reported for rats (Sapolsky, 1990; Sabatino et al., 1991) also could be involved, since elevated corticosterone is associated with decreased testosterone (Sapolsky, 1986). Finally, we mention studies showing that diverse pathological lesions are associated with GFAP mRNA elevations at all adult ages in mice (Goss et al., 1990); a role of decreased testosterone is implied, since gross lesions are associated with low testosterone (Nelson et al., 1975).

These hormonal and pathophysiological interactions give new candidates for resolving the causes of altered astrocyte activities with aging. We thus hope to extend observations relating age-related increases in hypothalamic astrocyte activities in the arcuate nucleus with the extent of exposure to ovarian estrogens (see above). So far in the hypothalamus, no estrogen-dependent neuron loss has been found (Miller et al., 1989). As mentioned above, the GnRH-containing neurons of the hypothalamus do not show changes in number during middle age, when the ovary-dependent neuroendocrine syndromes are first manifested (Hoffman and

Finch, 1986). These findings on astrocyte changes during aging in the hippocampus and hypothalamus maintain the possibility that neuron death or degeneration may not be the initiating event.

Nonhormonal interaction

On the other hand, we should consider the possibility that astrocytosis could be an initiating factor in neurodegeneration. Astrocytes are increasingly recognized for their complex functions and interactions with neurons. Neurons and astrocytes are of neuroectodermal origin; moreover, astrocytes, like neurons, have a higher inventory of complex RNA than other non-neural cell types (Kaplan et al., 1978). Neuronal migration along the paths formed by astrocytic processes is acknowledged as a fundamental process during development. Moreover, astrocytes from different brain regions influence the complexity of neurite branching (e.g., Chamak et al., 1987). Thus, we can consider how changes in astrocytic morphology and other functions could influence synaptic plasticity in the adult. Trophic interactions could be another aspect, since cultured astrocytes produce factors that enhance neuron survival (Kesslack et al., 1986). One astrocytic neurotrophin has been tentatively identified, NGF, which is secreted by C6 gliomas (Perez-Polo et al., 1977) and which also is found in astrocytes from developing brain (e.g., Lu et al., 1991). While the *in vivo* significance of astrocyte-derived trophic factors in the adult brain is unclear, it would be of interest to identify whether astrocytes cultured from rats of different ages showed changes in the secretion of NGF, other trophic factors, and various cytokines.

Last, we consider the role of astrocytes in relation to the neurodegeneration in AD and PD. In AD, astrocytes are clustered with microglia around amyloid deposits in senile plaques (e.g., Itagaki et al., 1989). Degenerating neurites are often seen in the vicinity of senile plaques, although degenerating neurites can be found elsewhere (Terry et al., 1991). The traditional explanation would be that astrocytes are activated by neurodegeneration and assist in the phagocytosis of cellular debris. However, in view of the activation of hippocampal astrocytes by reduction of testosterone as discussed above (Day et al., 1990) or by wasting conditions (Goss et al., 1990), we also could consider possible relationships to the reductions of gonadal steroids that would be typical in this age group (Finch, 1990; vom Saal and Finch, 1988) or to the wasting that is typical in the later stages of AD.

In PD, astrocytes are implicated from several lines of evidence: a relation to the neurotoxin MPTP and some incidental literature on astrocyte morphology and neurochemistry. In the now-famous example of parkinsonian syndromes induced by trace contaminants of MPTP in synthetic heroin, astrocytes are key to the production of the neurotoxin MPP^+, which kills substantia nigra neurons (reviewed in Poirier, 1987). The chemical modification of MPTP (1-methyl-4-phenyl-1,2,3,6-tetrahydropryidine) occurs in mitochondria of astrocytes through the activities of MAO B. The MPP^+ is then taken up by dopaminergic cells, which it can irreversibly damage. In view of the general trend for astrocytic activation during aging, it is particularly interesting that aging mice show greater susceptibility to MPTP (e.g., Gupta et al., 1986; Date et al., 1990; Langston et al., 1987). Whether this holds for other environmental toxins is an open question.

We also mention two little-cited studies that further implicate astrocytic involvement in PD. Issadorides (1971) examined neuro-vascular morphology in the substantia nigra compacta. Neurons from healthy age-matched individuals showed close contact with blood capillaries, but in PD the astrocytic processes appeared to expand the space between the remaining nigral neurons and capillaries. Other ultrastructural studies document that astrocytic processes are normally present on capillaries (e.g., the vascular feet observed in the rat dentate gyrus) (Kosaka and Hama, 1986). Thus, astrocytic morphological changes during aging could impair neuro-vascular contact. An early analysis of microdissected "neuroglia" (presumably astrocytes rather than oligodendroglia) from biopsy specimens of the globus pallidus indicated PD-associated shifts in base composition toward an enrichment of adenosine from 19% to 31% among the four RNA bases (Gomirato and Hydén, 1963; Hydén, 1967). This considerable increase of A (approximately 50%) is not easily accounted for as a relative increase of poly(A)RNA and implies major changes in the glial mRNA inventory.

Besides astrocytes, more recent studies show that microglia increases dramatically in the substantia nigra of PD and in many regions of AD, as judged by the immunopositivity for HLA-DR (McGeer et al., 1989; Styren et al., 1990). This glycoprotein indicates microglial activation and is one of the MHC type II glycoproteins also expressed on immunocompetent cells, where it presents foreign antigen to T-helper lymphocytes; the presence of HLA-DR on microglia does not establish that these macrophage-type cells are involved in a T-cell–mediated immune reaction in the brain. Because microglia can secrete TGF-β1 (e.g., in the AIDS-damaged brains) (Wahl et al., 1991) and because TGF-β1 can stimulate astrocytes (e.g., to produce GFAP as noted above), we suggest that astrocytic changes during AD and PD are driven by prior activation of microglia. Considerable increases of microglial numbers are reported in a sparse literature during aging in rodents (Vaughan and Peters, 1974), but the extent of microglial activation or of TGF-β production is unknown.

A working hypothesis is that astrocytic changes may be consequent to microglial activation during aging or other neurodegenerative conditions. After deafferenting lesions by perforant path transsection, hippocampal microglia are rapidly activated, before changes can be detected in astrocytes, in association with increased production of IL-1 (Fagan and Gage, 1990). In view of the ability of IL-1 to release TGF-β from astrocytes and microglia (da Cunha et al., 1990), a plausible sequence would be that IL-1 is a prime mover of increased TGF-β1 production and increased astrocytic activity during aging.

Summary

In conclusion, the age-related increase of astrocytes so far documented as a canonical change of aging in mice, rats, and humans could arise from several causes. The variety of factors and conditions that cause changes in astrocyte morphology suggests that the usual terms of astrocytic hyperactivity or astrocytosis are overly vague and must be further refined.

Astrocyte activities could change in response to degeneration of neurons or in response to sex steroids among other age-related endocrine changes. Such mechanisms would be examples of extrinsic causes in changed astrocyte activities during aging. It is also possible that astrocytes, as cells that are capable of dividing, manifest features of replicative senescence, such as observed for cultured diploid fibroblasts (the Hayflick phenomenon). This virtually unexplored issue could contribute an intrinsic dimension to age-related changes in astrocyte activities. The importance of astrocytes to many neural activities, including the production of neurotoxins and neurotrophins, suggests that, whatever the origins of the altered astrocytic activities during aging, there could be further consequences to neuronal function.

Acknowledgments. This work was supported by grants to CEF, AG-07909, and the John D. and Catherine T. MacArthur Foundation; to NJL, NRSA AG-05528. NRN is a Brookdale National Fellow.

References

Balcarek JM, Cowan NJ (1985): Structure of the mouse GFAP gene: implications for the evolution of the intermediate filament multigene family. *Nucleic Acids Res* 13:5527–5524.

Chamak B, Fellous A, Glowinski J, Prochiantz A (1987): MAP2 expression and neurite outgrowth and branching are coregulated through region-specific neuro-astroglial interactions. *J Neurosci* 7:3163–3170.

da Cunha A, Jefferson J, Jannotta FS, Vitkovic L (1990): Specificity of interleukin-1–mediated induction of transforming growth factor-β in glia. *Soc Neurosci Abst* vol. 16, #365.6.

Date I, Felten DL, Felten SY (1990): Long-term effect of MPTP in the mouse brain in relation to aging: neurochemical and imunocytochemical analysis. *Brain Res* 519:266–276.

Day JR, Laping NJ, McNeill TH, Schreiber SS, Pasinetti G, Finch CE (1990): Castration enhances expression of GFAP and SGP-2 in the intact and lesion-altered hippocampus of the adult male rat. *Mol Endocrinol* 4:1995–2002.

Fagan AM, Gage FH (1990): Cholinergic sprouting in the hippocampus: a proposed role for IL-1. *Exp Neurol* 110:105–120.

Finch CE (1990): *Longevity, Senescence, and the Genome.* Chicago: University of Chicago Press.

Finch CE, Felicio LS, Mobbs CV, Nelson JF (1984): Ovarian and steroidal influences on neuroendocrine aging processes in female rodents. *Endocrine Rev* 5:467–497.

Finch CE, Landfield PW (1985): Neuroendocrine and autonomic function in aging mammals. In: *Handbook of the Biology of Aging.* 2nd ed. Finch CE, Schneider EL, eds. New York: Van Nostrand.

Finch CE, Morgan DG (1990): RNA and protein metabolism in the aging brain. *Annu Rev Neurosci* 13:75–87.

Foster TC, Barnes CA, Rao G, McNaughton BL (1991): Increase in perforant path quantal size in aged F344 rats. *Neurobiol Aging* 12:441–448.

Geinisman Y, Bondareff W (1976): Decrease in the number of synapses in the senescent brain: a quantitative electron microscopic analysis of the dentate gyrus molecular layer in the rat. *Mech Aging Dev* 5:11–23.

Gomirato G, Hydén H (1963): A biochemical glia error in the Parkinson's disease. *Brain* 86:773–780.

Goss JR, Finch CE, Morgan DG (1990): GFAP RNA prevalence is increased in aging and in wasting mice. *Exp Neurol* 108:266–268.

Goss JR, Finch CE, Morgan DG (1991): Age-related changes in glial fibrillary acid protein mRNA in the mouse brain. *Neurobiol Aging* 12:165–170.

Gupta M, Gupta BK, Thomas R, Bruemmer V, Sladek JR Jr, Felton DL (1986): Aged mice are more sensitive to 1-methyl-4-phenyl-1,2,3,6-tetrahydropyridine than young adults. *Neurosc Lett* 70:326–331.

Hansen LA, Armstrong DM, Terry RD (1987): An immunohistochemical quantification of fibrous astrocyte in the aging human cerebral cortex. *Neurobiol Aging* 8:1–6.

Hoffman GE, Finch CE (1986): LHRH neurons in the female C57BL/6J mouse brain during reproductive aging: no loss up to middle-age. *Neurobiol Aging* 7:45–48.

Huang CJ, Laping NJ, Rozovsky I, Hogan TH, Finch CE (1993): Steroids differentially regulate the promoter activity of the rat GFAP gene to C6 glioma. *Endocrine Soc.* Abstract, in press.

Hydén H (1967): Dynamic aspects of the neuron-glia relationship. In: *The Neuron*, Hydén H, ed. Amsterdam: Elsevier.

Issidorides MR (1971): Neuronal vascular relationships in the zona compacta of normal and parkinsonian substantia nigra. *Brain Res* 25:289–299.

Itagaki S, McGeer PL, Akiyama H, Zhu S, Selkoe D (1989): Relationship of microglia and astrocytes to amyloid deposits of Alzheimer disease. *J Neuroimmunol* 24:173–182.

Johnson BR, Brinton RE (1989): steroid regulation of hippocampal nerve cell growth in culture. *Soc Neurosci Abstr* vol. 15, #427.14.

Jung-Testas I, Hu ZY, Baulieu EE, Robel P (1989): Neurosteroids: biosynthesis of pregnenolone and progesterone in primary cultures of rat glial cells. *Endocrinology* 125:2083–2091.

Kaplan BB, Schachter BS, Osterburg HH, de Vellis JS, Finch CE (1978): Sequence complexity of polyadenylated RNA obtained from rat brain regions and cultured rat cells of neural origin. *Biochemistry* 17:5516–5524.

Kohama SG, Anderson CP, Osterburg HH, May PC, Finch CE (1989): Oral administration of estradiol to young C57BL/6J mice induces age-like neuroendocrine dysfunctions in the regulation of estrous cycles. *Biol Reprod* 41:227–232.

Kohama SG, Brown SA, Finch CE, McNeill TH (1992): Chronic estradiol administration did not cause loss of hypothalamic LHRH or TIDA neurons in young or middle-aged C57BL/6J mice. *Brain Res* 574:341–345.

Kohama SG, Finch CE, McNeill TH (1991): Regional changes in GFAP mRNA across the estrous cycle. *Soc Neurosci Abst* vol. 17, #308.2.

Kosaka T, Hama K (1986): Three-dimensional structure of astrocytes from the rat dentate gyrus. *J Comp Neurol* 249:242–260.

Landfield PW, Waymire J, Lynch G (1978): Hippocampal aging and adrenocorticoids: quantitative correlations. *Science* 202:1098–1102.

Langston JW, Irwin I, DeLanney LE (1987): Molecular mechanisms of MPTP-induced toxicity. VII. The biotransformation of MPTP and disposition of MPP^+: the effects of aging. *Life Sci* 40:749–754.

Laping NJ, Nichols NR, Day JR, Finch CE (1991): Corticosterone differentially regulates the bilateral response of astrocyte mRNAs to entorhinal cortex lesions in male rats. *Mol Brain Res* 10:291–297.

Lindholm D, Hengerer B, Zafra F, Thoenen H (1990): Transforming growth factor-β1 stimulates expression of nerver growth factor in the rat CNS. *NeuroReport* 1:9–12.

Lu B, Yokoyama M, Dreyfus CF, Black IB (1991): NGF gene expression in actively growing brain glia. *J Neurosci* 11:318–326.

McGeer PL, Akiyama H, Itagaki S, McGeer EG (1989): Immune system response in Alzheimer's disease. *Can J Neurol Sci* 16:516–527.

Mobbs CV, Finch CE (1992): Estrogen-induced neuroendocrine impairment as a mechanism of senescence in female C57BL/6J mice. *J Gerontol* 47:B48–51.

Nakatani Y, Horikoshi M, Brenner M, Yamamoto T, Besnard F, Roeder RG, Freese E (1990): A downstream initiation element required for efficient TATA box binding and in vitro function of TFII. *Nature* 348:86–88.

Nelson FJ, Latham KR, Finch CE (1975): Plasma testosterone levels in C57BL/6J male mice: effects of age and disease. *Acta Endocrinol* 80:744–752.

Nichols NR, Laping NJ, Day JR, Finch CE (1991): Increases in transforming growth factor-β mRNA in hippocampus during response to entorhinal cortex lesions in intact and adrenalectomized rats. *J Neurosci Res* 28:134–139.

Nichols NR, Osterburg HH, Masters JN, Millar SL, Finch CE (1990): Messenger RNA for glial fibrillary acidic protein is decreased in rat brain following acute and chronic corticosterone treatment. *Mol Brain Res* 7:1–7.

Norenburg MD (1987): The role of astrocytes in hepatic encephalopathy. *Neurochem Pathol* 6:13–33.

O'Callaghan JP, Miller DB (1991): The concentration of glial fibrillary acidic protein increases with age in the mouse and rat brain. *Neurobiol Aging* 12:171–174.

Perez-Polo JR, Hall K, Livingstone K, Westlund K (1977): steroid induction of nerve growth factor synthesis in cell culture. *Life Sci* 21:1535–1545.

Poirier J (1987): Pathophysiology and biochemical mechanisms involved in MPTP-induced parkinsonism. *J Am Geriatr Soc* 35:660–668.

Poirier J, Hess M, May PC, Finch CE (1991): Astrocytic apolipoprotein E mRNA and GFAP mRNA in hippocampus after entorhinal cortex lesioning. *Mol Brain Res* 11:97–106.

Sabatino F. Masoro EJ, McMahan CA, Kuhn RW (1991): Assessment of the role of the glucocorticoid system in aging processes and in the action of food restriction. *J Gerontol* B171–179.

Sakai Y, Rawson C, Lindberg K, Barnes D (1990): Serum and TGF-β regulate GFAP in serum-free derived mouse embryo cells. *Proc Nat Acad Sci* 87:8387–8382.

Sapolsky RM (1986): Stress-induced elevation of testosterone concentrations in high-ranking baboons: role of catecholamines. *Endocrinology* 118:1630–1635.

Sapolsky RM (1990): The adrenalcortical axis. In: *Handbook of the Biology of Aging*, 3rd ed, Schneider EL, Rowe JW, eds. New York: Academic Press.

Schipper H, Brawer JR, Nelson JF, Felicio LS, Finch CE (1981): Role of the gonads in the histologic aging of the hypothalamic arcuate nucleus. *Biol Reprod* 25:413–419.

Simmerly RB, Chang C, Muramatsu M, Swanson LW (1990): Distribution of androgen and estrogen receptor mRNA-containing cells in the rat brain: an in situ hybridization study. *J Comp Neurol* 294:76–95.

Styren SD, Civin WH, Rogers J (1990): Molecular, cellular, and pathological characterization of HLA-DR immunoreactivity in normal elderly and Alzheimer's disease brain. *Exp Neurol* 110: 93–104.

Vaughan DW, Peters A (1974): Neuroglial cells in the cerebral cortex of rats from young adulthood to old age: an electron microscope study. *J Neurocytol* 3:405–429.

vom Saal FS, Finch CE (1988): Reproductive senescence: phenomena and mechanisms in mammals and selected vertebrates. In: *Physiology of Reproduction*, vol. 2, Knobil E, ed. New York: Raven Press.

Wahl SM, Allen JB, McCartney-Francis N, Morganti-Kossman MC, Kossman T, Ellingsworth L, Mai UEH, Mergenhagen SE, Orenstein JM (1991): Macrophage and astrocyte-derived transforming growth factor β as a mediator of central nervous system dysfunction in acquired immune deficiency syndrome. *J Exp Med* 73:981–991.

Chapter 3

The Natural History of Aging

K. W. Lange and P. Riederer

Aging can be defined as deleterious alterations reducing the likelihood of an individual's survival (Comfort, 1964) and is characterized by increasing deviations from an ideal functional state. Several biological theories have been put forward to explain these alterations. Most of these theories assume that the life span of a species is genetically determined (Shock, 1977). The hypothesis of a genetic basis of aging is supported by studies showing that many types of diploid body cells such as cultured fibroblasts can divide only a limited number of times (Hayflick, 1965). This theory may not apply to the brain, since neurons are postmitotic and glial cells divide infrequently. Genetic cellular theories suggest that aging is a result of damage to the genetic information determining cellular protein formation. The DNA replication theory assumes that aging may result from a gradual accumulation of random errors with eventual functional or reproductive death of individual cells. The genetic information may be altered by damage to the DNA or by faulty transmission of genetic information by RNA. Should DNA be damaged by mutations at a faster rate than it can be repaired, nondividing cells such as neurons would be particularly affected (Sinex, 1974). Nongenetic cellular theories emphasize alterations occurring after formation of cellular proteins. The accumulation of harmful substances such as lipofuscin (Carpenter, 1965), the formation of cross-linkages between components of the same molecule or between different molecules (Bjorksten, 1968), and free radical formation (Harman, 1968) may interfere with cellular metabolism and cause cell dysfunction and eventual cell death. It is difficult to evaluate these theories for aging in the central nervous system, since there are no generally accepted age-related alterations in brain function. These theo-

ADVANCES IN RESEARCH ON NEURODEGENERATION, I
D.B. Calne et al.

ries may apply indirectly to the central nervous system, since the brain interacts with and depends on various other body systems. This view is expressed by theories of aging focusing on the inability of the endocrine (Finch, 1976) and immune (Walford, 1969) systems to coordinate bodily functions adequately.

Research on aging has emphasized age-related functional losses. Major effects of age have been shown in regard to vision, hearing, glucose tolerance, renal function, bone density, activity of the sympathetic nervous system, endocrine function, immune function, cognition, and behavioral functions. Some aged individuals fall within the normal range for adults while others present with major deficits. Loss of function may be due to illness or environmental factors rather than aging itself. Recent neurobiological literature suggests that more systems and functions are preserved than may be predicted from the hypothesis that aging is associated with a continuous and inevitable decline until death (Table 1).

Table 1. Neuronal Loss in Man During Aging

Brain regions with age-related neuronal loss
Large neurons in frontal, temporal, and parietal cortex
Substantia nigra pars compacta
Locus ceruleus
Cerebellum
Brain regions without age-related neuronal loss
Total number of neurons in frontal, temporal, and parietal cortex
Pyramidal neurons in hippocampus
Amygdala
Putamen
Raphe nuclei
Nucleus basalis of Meynert
Supraoptic and paraventricular nuclei of the hypothalamus
Inferior olivary nucleus
Cranial nerve nuclei IV, VI, VII
Ventral cochlear nucleus

See Coleman and Flood (1987) and Jellinger (1992).

Brain Changes Associated with Aging

The brain decreases in volume and weight during senescence. Alterations are minimal up to the age of 45 years, after which time a significant decline in brain weight can be observed. The brain weight of elderly humans is 7% to 8% less than that of the peak of adult brain weight

(Creasey and Rapaport, 1985). There is, however, no diffuse loss of neurons or other brain cells, but brain cell atrophy shows great variability in various brain regions. While some regions of the cortex lose as much as 50% of their cells, other areas do not appear to be affected (Henderson et al., 1980). Loss of neural cells has also been reported for the basal ganglia (McGeer et al., 1977) and the hippocampus (Henderson et al., 1980) whereas other subcortical areas seem to be spared (Konigsmark and Murphy, 1970). The substantia nigra in humans shows a progressive loss of dopaminergic neurons, beginning at approximately 30 years of age (McGeer et al., 1977). Cell loss in the substantia nigra of aged humans can exceed 60%, whereas it is less than 20% in old rodents (Morgan et al., 1987).

Despite neural cell losses during senescence, the functional capacity of neuronal circuits in the brain may be preserved better than would be predicted from cell loss alone. Reactive synaptogenesis has been reported in the injured aged brain (Cotman and Anderson, 1988) and neuronal loss during aging may trigger the surviving neurons to sprout and replace lost connections. In brain areas showing partial neuronal loss, dendritic arbors continue to grow between middle and old age. Dentate granule cells in humans, for example, show an increase in dendritic length per cell with increasing age, and only in very old age does the dendritic tree appear to regress to that of mature younger adults (Coleman and Flood, 1986). Compensatory mechanisms of remodeling and growth therefore may be able to maintain and adapt brain function during senescence.

Aging and Motor Function

It is generally accepted that marked alterations in motor activity and function occur with aging. Both simple and complex motor tasks show a decline in speed over the life span. This decline arises from alterations both in the muscle system and in central information-processing networks that command motor functions. The physiological substrates of these motor declines are evident at many levels of the neuromotor system. Both morphological and neurochemical alterations have been reported in the brain. In humans neuronal loss appears to be most pronounced in the cerebellum (Hall et al., 1975) and in one of its major sources of afferents, the locus ceruleus (Vijayashankar and Brody, 1979; Wree et al., 1980). In contrast to the locus ceruleus, no decrease in absolute cell numbers with aging has been found in the inferior olive (Monagle and

Brody, 1974), ventral cochlear nucleus (Konigsmark and Murphy, 1972), facial nucleus (Maleci, 1934), nuclei of the trochlear and abducens nerves (Vijayashankar and Brody, 1977a,b), or mamillary bodies (Wilkinson and Davies, 1978). Neuron density appears to decline markedly in the human cerebral cortex (Brody 1955, 1970; Henderson et al., 1980). The study of dendrites in the aging brain has drawn particular attention to the loss of Betz cells in the human motor cortex (Scheibel et al., 1977). Ventral horn motor neurons, the final common pathway of the somatic motor system, have also been shown to undergo degenerative changes with advancing age (Scheibel, 1979).

The frequent occurrence of extrapyramidal symptoms in aged humans (Kane et al., 1982) suggests that the nigrostriatal dopaminergic system degenerates during aging. The number of neurons in the substantia nigra decreases age-dependently in both animals and humans (McGeer et al., 1977; Gibb and Lees, 1991) and the concentration of striatal dopamine declines during aging (Adolfsson et al., 1979; Riederer and Wuketich, 1976). The increased striatal dopamine turnover, as expressed by the ratio of homovanillic acid to dopamine, may compensate adequently for the dopamine loss and may maintain normal motor functions of the elderly, but may be unable to cope with increased functional requirements. Other age-dependent alterations concerning dopaminergic systems in the brain have been reported for dopamine synthesis and metabolism (McNamara et al., 1974), dopamine uptake (Jonec and Finch, 1975), and dopamine-sensitive adenylate-cyclase activity (Govoni et al., 1977; Schmidt and Thornberry, 1978). A loss of striatal dopamine D_2 receptors has been shown in rodents, primates, and man (for review see Lai et al., 1987). Contradictory results, however, have been reported on the age-related changes in the density of dopamine D_1 receptors. In humans an increase (Morgan et al., 1987), no change (Rinne, 1987), or a decrease (Seeman et al., 1987; Rinne et al., 1990) have been found. The discrepant results concerning striatal D_1 receptor alterations with aging may be explained by different ligands used for receptor binding. All studies using the highly selective D_1 ligand [^{3}H]-SCH 23390 have shown an age-dependent decline of D_1 receptor concentrations in the striatum (for review see Lange and Kalveram, 1991). The ratio of D_1 to D_2 receptors has been reported to remain constant during the course of aging in humans (Rinne et al., 1990) and rats (Lange and Kalveram, 1991). Another study, however, showed an age-related decline in the D_1:D_2 ratio; that is, more D_1 receptors than D_2 receptors are lost, leading to an imbalance between the two receptor subtypes and to possible deficits in motor behavior. The *in vivo* assessment of striatal dopaminergic function in humans with positron emission tomography (PET) has produced conflicting results. Both cross-sectional (Martin et al., 1989) and longitudinal (Bhatt et al., 1991) PET studies using the tracer [^{18}F]-fluorodopa showed an age-related decline in dopaminergic function similar to the reduction in substantia nigra cells reported by McGeer et al., (1977). Another cross-sectional PET study was unable to find a reduction during the course of aging (Sawle et al., 1990). This study may have included a relatively high proportion of young persons with lower levels of dopaminergic function and a possibly lower life expectancy. To reduce genetic heterogeneity with regard to longevity, both young individuals and their grandparents were examined with PET. Preliminary results point to a decrease in striatal [^{18}F]-fluorodopa within each grandchild–grandparent pair and indicate that a decline in striatal dopaminergic function occurs with aging (Snow and Cordes, 1992).

Aging and Cognitive Function

Early studies on the development of intelligence with aging concluded that a general decline in cognitive functions was inevitable as a result of biological aging processes (Wechsler, 1958). Problems regarding the interpretation of the results of psychometric studies on aging arise from the use of either cross-sectional or longitudinal studies. An instructive series of cross-sequential studies on successive cohorts of individuals across the adult age range allows a comparison of cross-sectional and longitudinal data from the same study populations (Schaie and Labouvie-Vief, 1974). Cross-sectional comparisons between age groups for individuals tested in 1963 showed lower scores in many cognitive faculties for the older groups, a finding commonly interpreted as reflecting a decline in performance with advancing age. When the same subjects were retested in 1970, however, they did not show the age-related decline that would have been predicted on the basis of the earlier cross-sectional comparisons. The longitudinal data suggested that deterioration of performance in both verbal tests (crystallized intelligence) and tasks testing reasoning and spatial abilities (fluid intelligence) occurred at substantially later ages than the cross-sectional comparisons had implied. Differences between longitudinal and cross-sectional studies exemplify a cohort effect in which successive age cohorts show different test results at the same chronological age. Such differences cannot be explained by factors intrinsic to aging that are presumably constant between age cohorts. Cohort differences reflect rather extrinsic factors such as differing education or nutrition. The influence of education on test performance commonly considered to be age-related has been demonstrated by an analysis of scores in the Wechsler Adult Intelligence Scale (Green, 1969). Comparisons across age groups in a stratified random sample demonstrated clear age-dependent differences. Similar comparisons performed between age groups limited to persons of equal education, however, showed that the influence of age became insignificant. These findings indicate that the decline in cognitive functions considered to be the consequence of aging per se is at least partly caused by extrinsic factors.

Studies of age-related alterations of cognition must distinguish between changes in cognitive activity itself and those secondary to deterioration in the peripheral sensory mechanism. Cognitive functions may be impaired solely by sensory disorders of vision or hearing. This view is supported by studies investigating cerebral energy metabolism such as cerebral blood flow, cerebral metabolic rate for oxygen, and cerebral metabolic rate for glucose in aged animals and humans. Rates of glucose utilization have been shown to decline only slightly in old rats that have only a few age-related neuropathological alterations (Smith et al., 1980). These alterations were found in nearly all structures of the visual and auditory systems. In the limbic and sensorimotor systems only a few regions were affected by aging; notably the striatum and the globus pallidus. The reductions in glucose utilization in the visual and auditory systems may reflect degenerative alterations in the retina and cochlea. The changes in the striatum and globus pallidus reflect the fact that the extrapyramidal motor system is a prime site of senescent alterations. No effect of age on local rates of glucose utilization in 30 brain regions or on the average rate of glucose consumption for the entire brain could be found in healthy men between the ages of 21 and 83 (Duara et al., 1983, 1984). These men were examined under conditions of reduced visual and auditory stimulation. Another study of individuals between 18 and 78 years of age without sensory deprivation showed an age-dependent decline in glucose utilization in both the brain as a whole and in all brain areas examined. These results of unchanged cerebral glucose utilization in the sensory-deprived state and decreased rate of glucose consumption under conditions of sensory input suggested that the input and/or the processing of sensory information is affected in the aged. Many of the alterations in cerebral metabolism may reflect a decrease in functional activity due to a partial deafferentiation of sensory systems. Continued sensory input and continued use of the brain's functional capacity may be necessary for maintenance of an optimally responsive brain in old age. Decreased sensory input to the central nervous system due to degenerative alterations in the sensory organs may cause reductions in metabolism and blood flow as well as morphological involution in the brain regions involved in processing sensory information.

Neurodegenerative diseases such as Alzheimer's disease and Parkinson's disease are associated with neuronal loss in the nucleus basalis of Meynert (Rossor et al., 1982; Candy et al., 1983; Whitehouse et al., 1983). The loss of cortical choline acetyltransferase activity in patients with Alzheimer's disease or Parkinson's disease has been reported to correlate with the severity of dementia (Perry et al., 1978, 1985; Lange et al., 1993). Whether or not an age-related neuronal loss in the nucleus basalis of Meynert occurs in persons with no neuropsychiatric disease is unclear. Although a decline in cell number over a normal life span has been reported by one group (McGeer et al., 1984), this change could not be confirmed by other groups (Chui et al., 1984; Vogels et al., 1990).

Implications of Age-Related Neuronal Changes

There is evidence that a continuous decline in motor abilities occurs during aging. The nigrostriatal system appears to be one of the prime sites associated with age-related neuronal losses. Both neuropsychological and neurobiological studies suggest that cognitive functions and their underlying brain systems are preserved much longer in aged individuals not affected by neurodegenerative disorders such as Alzheimer's disease. It may therefore be possible that parts of the motor system such as the substantia nigra play a key role as pacemaker in the rate of aging. These brain structures may be programmed to signal the passage of one stage of life span to the next. Alternatively, they may be programmed only until sexual maturation has been achieved and continuation of the species guaranteed. In this case aging may result from a lack of programming and an imbalance of neural and endocrine signals and their interactions. The relative preservation of cognitive functions in aged individuals may reflect a survival advantage in evolutionary terms. In most animal species, biological advantage can be measured by the ability to reach reproductive age and by the number of offspring produced. In humans, however, biological advantage may be associated with successful aging so that the process of acquisition and transmission of information can be maximized.

It seems paradoxical that the brain as the principle organ for the accumulation and transmission of knowledge is subject to age-related structural damage in the form of neuronal attrition. This paradox is based on the assumption that a greater number of neurons, is better with regard to function. Although in brain areas such as the striatum a certain degree of dopamine loss causes impaired function, at present there are no data available showing that neuronal attrition is necessarily the structural substrate for age-related cognitive decline. Since cortical neuronal densities undergo a substantial decline not only in senescence but also from childhood to adulthood (Brody, 1955), it is simplistic to expect a simple correlation between neuronal number and behavior. Neuronal attrition may not therefore be synonymous with cognitive decline. The loss of neurons and synapses at certain stages of an individual's development could represent a process of gradual fine-tuning establishing more effective patterns of information processing and behavior and replacing the global and diffuse activation patterns of earlier developmental phases. At present it appears to be impossible to express the "wisdom" of old age in terms of neuronal counts or other neurobiological substrates.

References

Adolfsson R, Gottfries C-G, Roos B-E, Winblad B (1979): Post-mortem distribution of dopamine related to age, and a review of the literature. *J Neural Transm* 45:81–105.

Bhatt M, Snow BJ, Martin WRW, Pate BD, Ruth TJ, Calne DB (1991): Positron emission tomography suggests that the rate of progression of idiopathic parkinsonism is slow. *Ann Neurol* 29:673–677.

Bjorksten J (1968): The cross linkage theory of aging. *J Am Geriatr Soc* 16:408–427.

Brody H (1955): Organization of the cerebral cortex III. A study of aging in the human cerebral cortex. *J Comp Neurol* 102:511–556.

Candy JM, Perry RH, Perry EK, Irving D, Blessed G, Fairbairn AF, Tomlinson BE (1983): Pathological changes in the nucleus of Meynert in Alzheimer's and Parkinson's diseases. *J Neurol Sci* 54:277–289.

Carpenter DG (1965): Diffusion theory of aging. *J Gerontol* 20:191–195.

Chui HC, Bondareff W, Zarow C, Slager U (1984): Stability of neuronal number in the human nucleus basalis of Meynert with age. *Neurobiol Aging* 5:83–88.

Coleman PD, Flood DG (1986): Dendritic proliferation in the aging brain as a compensatory repair mechanism. In: *Progress in Brain Research, Vol. 70*, Swaab DF, Fliers E, Mirmiran M, Van Gool WA, Van Haaren F, eds. Amsterdam: Elsevier.

Coleman PD, Flood DG (1987): Neuron numbers and dendritic extent in normal aging and Alzheimer's disease. *Neurobiol Aging* 8:521–545.

Comfort A (1964): *Aging: The Biology of Senescence.* New York: Holt, Rinehart and Winston.

Cotman CW, Anderson KJ (1988): Synaptic plasticity and functional stabilization in the hippocampal formation: possible role in Alzheimer's disease. In: *Physiologic Basis for Functional Recovery in Neurological Disease*, Waxman ZS, ed. New York: Raven Press.

Creasey H, Rapoport SI (1985): The aging human brain. *Ann Neurol* 17:2–10.

Duara R, Margolin RA, Robertson-Tchabo EA, London ED, Schwartz M, Renfrew JW, Koziarz BJ, Sundaram M, Grady C, Moore AM, Ingvar DH, Sokoloff L, Weingartner H, Kessler RM, Manning RG, Channing MA, Cutler NR, Rapoport SI (1983): Cerebral glucose utilization, as measured with positron emission tomography in 21 resting healthy men between the ages of 21 and 83 years. *Brain* 106:761–775.

Duara R, Grady C, Haxby J, Ingvar D, Sokoloff L, Margolin RA, Manning RG, Cutler NR, Rapoport SI (1984): Human brain glucose utilization and cognitive function in relation to age. *Ann Neurol* 16:702–713.

Finch CE (1976): The regulation of physiological changes during mammalian aging. *Q Rev Biol* 51:49–83.

Gibb WRG, Lees AJ (1991): Anatomy, pigmentation, ventral and dorsal subpopulations of the substantia nigra, and differential cell death in Parkinson's disease. *J Neurol Neurosurg Psychiatr* 54:388–396.

Govoni S, Loddo P, Spano PF, Trabucchi (1977): Dopamine receptor sensitivity in brain and retina of rats during aging. *Brain Res* 138:565–570.

Green RF (1969): Age-intelligence relationship between ages sixteen and sixty four. *Dev Psychol* 1:618L-627.

Hall TC, Miller AKH, Corsellis JAN (1975): Variations in the human Purkinje cell population according to age and sex. *Neuropathol Appl Neurobiol* 1:267–292.

Harman D (1968): Free radical theory of aging: Effect of free radical reaction inhibitors on the mortality rate of male LAF_1 mice. *J Gerontol* 23:476–482.

Hayflick L (1965): The limited *in vitro* lifetime of human diploid cell strains. *Exp Cell Res* 37:614–636.

Henderson G, Tomlinson BE, Gibson PH (1980): Cell counts in human cerebral cortex in normal adults throughout life using an image analysing computer. *J Neurol Sci* 46:113–136.

Jellinger KA (1992): Morphology of the aging brain and relations to Alzheimer's disease (this volume).

Jonec V, Finch CE (1975): Ageing and dopamine uptake by subcellular fractions of the C57BL/6J male mouse brain. *Brain Res* 91:197–215.

Kane JM, Weinhold P, Kinon B, Wegner J, Leader M (1982): Prevalance of abnormal involuntary movements ("spontaneous dyskinesias") in the normal elderly. *Psychopharmacology* 77: 105–108.

Konigsmark BM, Murphy EA (1970): Neuronal population in the human brain. *Nature* 228:1225–1336.

Konigsmark BM, Murphy EA (1972): Volume of the ventral cochlear nucleus in man: Its relationship to neuronal population and age. *J Neuropathol Exp Neurol* 31:304–316.

Lai H, Bowden DM, Horita A (1987): Age-related decreases in dopamine receptors in the caudate nucleus and putamen of the rhesus monkey (macaca mulatta). *Neurobiol Aging* 8:45–49.

Lange KW, Wells FR, Jenner P, Marsden CD (1993): Altered muscarinic and nicotinic receptor densities in cortical and subcortical brain regions in Parkinson's disease. *J Neurochem* 60:107–203.

Lange KW, Kalveram KT (1991): Effects of ageing on dopamine D-1 and D-2 receptors in the striatum of the rat. *Med. Sci. Res.* 19:77–78.

Maleci O (1934): Contributo all conoscenza delle variazoni quantitative delle cellule nervose nelle senescenza. *Arch Ital Anat* 33:883–901.

Martin WRW, Palmer MR, Patlak CS, Calne DB (1989): Nigrostriatal function in man studied with postron emission tomography. *Ann Neurol* 26:535–542.

McGeer PL, McGeer EG, Suzuki JS (1977): Aging and extrapyramidal function. *Arch Neurol* 34:33–35.

McGeer PL, McGeer EG, Suzuki J, Doman CE, Nagai T (1984): Aging, Alzheimer's disease, and the cholinergic system of the basal forebrain. *Neurology* 34:743–745.

McNamara MC, Miller AT, Vernon AB, David JN (1974): Age-related changes in the effect of electroconvulsive shock (ECS) on the in vivo hydroxylation of tyrosine and tryptophan in rat brain. *Brain Res* 131:313–320.

Monagle RD, Brody H (1974): The effects of age upon the main nucleus of the inferior olive in the human. *J Comp Neurol* 155:61–66.

Morgan DG, Marcusson JO, Nyberg P, Wester P, Winblad B, Gordon MN, Finch CE (1987): Divergent changes in D-1 and D-2 dopamine binding sites in human brain during aging. *Neurobiol Aging* 8:195–201.

Perry EK, Tomlinson BE, Blessed G, Bergmann K, Gibson PH, Perry RH (1978): Correlation of cholinergic abnormalities with senile plaques and mental test scores in senile dementia. *Br Med J* ii:1457–1459.

Perry EK, Curtis M, Dick DJ, Candy JM, Atack JR, Bloxham CA, Blessed G, Fairbairn A, Tomlinson BE, Perry EH (1985): Cholinergic correlates of cognitive impairment in Parkinson's disease: comparison with Alzheimer's disease. *J Neurol Neurosurg Psychiat* 48:413–421.

Riederer P, Wuketich S (1976): Time course of nigrostriatal degeneration in Parkinson's disease. *J Neural Transm* 38:277–301.

Rinne JO (1987): Muscarinic and dopaminergic receptors in the aging human brain. *Brain Res* 404:162–168.

Rinne JO, Lönnberg P, Marjamäki P (1990): Age-dependent decline in human brain dopamine D_1 and D_2 receptors. *Brain Res.* 508:349–352.

Rossor MN, Garrett NJ, Johnson AL, Mountjoy CQ, Roth M, Iversen LL (1982): A post-mortem study of the cholinergic and GABA systems in senile dementia. *Brain* 105:313–330.

Sawle GV, Colebatch JG, Shah A, Brooks DJ, Marsden CD, Frackowiak RSJ (1990): Striatal function in normal aging: implications for Parkinson's disease. *Ann Neurol* 28:799–804.

Schaie KW, Labouvie-Vief G (1974): Generational and cohort-specific differences in adult cognitive behavior: a fourteen-year cross-sequential study. *Dev Psychol* 10:305–320.

Scheibel AB (1979): Aging in human motor control systems. *Aging* 10:297–310.

Scheibel ME, Tomiyasu U, Scheibel AB (1977): The aging human Betz cell. *Exp Neurol* 56:598–609.

Schmidt MJ, Thornberry JF (1978): Cyclic AMP and cyclic GMP accumulation in vitro in brain regions of young, old and aged rats. *Brain Res* 139:169–177.

Seeman P, Bzowej NH, Guan HC, Bergeron C, Becker LE, Reynolds GP, Bird ED, Riederer P, Jellinger K, Watanabe S, Tourtellotte WW (1987): Human brain dopamine receptors in children and aging adults. *Synapse* 1:399–404.

Shock NW (1977): Biological theories of aging. In: *Handbook of the Psychology of Aging*. Schaie KW, ed. New York: Van Nostrand Reinhold.

Sinex FM (1974): The mutation theory of aging. In: *Theoretical Aspects of Aging*. Rockstein M, ed. New York: Academic Press.

Smith CB, Goochee C, Rapoport SI, Sokoloff L (1980): Effects of ageing on local rates of cerebral glucose utilization in the rat. *Brain* 103:361–365.

Snow B, Cordes M (1992): Definition of natural aging (this volume).

Vijayashankar N, Brody H (1977a): Aging in the human brain stem. A Study of the nucleus of the trochlear nerve. *Acta Anat* 99:169–172.

Vijayashankar N, Brody H (1977b): A quantitative study of the pigmented neurons in the nuclei locus coeruleus and subcoeruleus in man as related to aging. *Int Neuropathol Exp Neurol* 38:490–497.

Vogels OJM, Broere AJ, Ter Laak HJ, Ten Doncklar HJ, Nieuwenhuis R, Schulte BPM (1990): Cell loss and shrinkage in the nucleus basalis Meynert complex in Alzheimer's disease. *Neurobiol Aging* 11:3–13.

Walford RL (1969): *The Immunologic Theory of Aging*. Baltimore: Williams & Wilkins.

Wechsler D (1958): *The Measurement and Appraisal of Adult Intelligence (4th ed)*. Baltimore: Williams & Wilkins.

Whitehouse PJ, Hedreen JC, White III CL, Price DL (1983): Basal forebrain neurons in dementia of Parkinson's disease. *Ann Neurol* 13:243–248.

Wilkinson A, Davies I (1978): The influence of age and dementia on the neurone population of the mammilary bodies. *Age Ageing* 7:151–160.

Wree A, Braak H, Schleicher A, Zilles K (1980): Biomathematical analysis of the neuronal loss in the aging human brain of both sexes demonstrated in pigment preparations of the pars cerebellaris locus coerulei. *Anat Embryol* 106:105–119.

Part II

Parkinson's Disease

Chapter 4

Normality, Disturbances, and Diseases

Donald B. Calne

Terminology

The major neurodegenerative diseases, Alzheimer's disease, Parkinson's disease, and motoneuron disease, share many features. They are inexorably progressive, they predominate in later life, familial variants exist with younger onset, and underlying each is a focus of neuronal death in a functionally related system of nerve cells that display cytoplasmic inclusion bodies containing cytoskeletal products.

A satisfactory definition has eluded Alzheimer's disease, Parkinson's disease, and motoneuron disease. This difficulty in determining diagnostic criteria is a further important shared feature, which derives, at least in part, from the similarity between the normal aging process in the central nervous system and the findings in mild degenerative disease. An important component of the problem is the task of drawing a line between normality and disease. This problem is compounded by the notion that many asymptomatic normal elderly persons may have subclinical stigmata of neurodegenerative disease because the evolution of their pathology is likely to be very slow.

One approach to these various dilemmas is to formulate working definitions not only for the diseases, but also for the terminology employed. I would suggest the following.

Normality

Normal subjects are average; they conform with the majority of members of a population. Normal attributes are quantifiable and are scattered

ADVANCES IN RESEARCH ON NEURODEGENERATION, I
D.B. Calne et al.

around a mean value for whatever clinical feature, physiological function, or morphological measure may be under consideration. In defining control populations for study, subjects must be matched for age, sex, and race. In particular, it is important to avoid selecting "super normals" by bias in the choice of controls. "Super normals" are of interest, primarily, for comparison with average (or true) normals rather than for comparison with disturbed subjects.

Disturbance

Disturbance is a useful term for any clinical physiological or morphological index that differs from control (average) values by an arbitrary amount, for example as two standard errors of the mean. By analogy with studies on blood pressure or blood glucose, it is useful to formulate criteria for defining abnormalities that are disturbances, but that are not sufficient to generate symptoms. In the context of Parkinson's disease, an asymptomatic subject would have a disturbance if, for example, the rate of eyeblinking was significantly reduced, or post mortem findings revealed a significant decrease in the number of nerve cells in the zona compacta of the substantia nigra. Like normality, the term disturbance assumes precise meaning if it is defined quantitatively.

Disease

Disease is defined by the presence of symptoms (Calne and Calne, 1988). Ideally, there should be an associated cause, pathological picture, natural history, and response to treatment. Where it is not possible to assemble the full profile of a disease, it is often preferable to employ the term *syndrome*, for a constellation of symptoms and signs that are associated (literally run together).

With this background of terminology, we have recently proposed a clinical definition of Parkinson's disease (Takahashi and Calne, 1992). The criteria—both for inclusion and exclusion—are summarized in Table 1. We also have suggested a series of clinically defined levels of certainty for making a diagnosis (Calne, Snow, and Lee, 1992). These are summarized in Table 2.

Table 1. Diagnostic Criteria

Clinical Parkinson's disease (idiopathic parkinsonism) comprises at least two features out of the triad—tremor, rigidity, and bradykinesia—with the following inclusion and exclusion criteria:

1. No detectable cause
2. No cerebellar deficits
3. Pyramidal features limited to possible hyperreflexia and an extensor plantar response
4. No evidence of lower motoneuron dysfunction
5. Gaze deficits limited to possible paresis of upward conjugate gaze
6. Autonomic deficits limited to minor dysfunction and insufficient to cause repeated syncope; in this context, drug-induced autonomic failure must be excluded

Table 2. Stages of Confidence in Diagnosing Idiopathic Parkinsonism (IP)

Categories of IP:

- Clinically possible
- Clinically probable
- Clinical definite

\+ Laboratory support for any of the above

Clinically possible IP:
Any one of:

- resting tremor
- rigidity
- bradykinesia

Clinically probable IP:
Any two of:

- resting tremor
- rigidity
- bradykinesia
- impaired postural reflexes

or
Any one of the above first three displaying asymmetry

Clinically definite IP:
Any three of:

- tremor
- rigidity
- bradykinesia
- impaired postural reflexes

or
Any two of the above with one of the first three displaying asymmetry

Laboratory-supported IP

Positive investigations (notably PET) with clinically possible, probable, or definite IP

Conclusions

The study of neurodegenerative diseases is seriously compromised by confusion over how to define normality and what constitutes each disease. How can mild disease states be distinguished from normal aging? How do we interpret abnormal function or structure in the absence of symptoms?

Another major problem is the lack of specificity of so many clinical and pathological findings. For example, parkinsonism is now recognized as a syndrome for which there are many causes, such as progressive supranuclear palsy or exposure to neuroleptics. Similarly, Lewy bodies are currently being found in many patients with Alzheimer's disease, in addition to Parkinson's disease.

As an initial step toward resolving the confusion, it may be helpful, in the study of neurodegenerative diseases, if we adopt separate clinical and pathological criteria for diagnosis. As a pragmatic tool for facilitating research, there may also be merit in defining levels of confidence: possible, probable, and definite. We have formulated such a scheme for the clinical diagnosis of Parkinson's disease (Calne et al., 1991). It may be possible to establish similar levels of confidence for the pathological diagnosis. The same principles may be applied to motoneuron disease and Alzheimer's disease.

References

Calne DB, Calne JS (1988): Normality and disease. *Can J Neurol Sci* 15:3–4.

Calne DB, Snow BJ, Lee C (1991): *Criteria for diagnosing Parkinson's disease. Proceedings of the Iron Meeting. Ann Neurol* 32:S125–127.

Takahashi H Calne DB (1993): *What is Parkinson's disease. Proceedings of the European Conference on Parkinson's disease and extrapyramidal disorders.* Como, Italy: John Libbey CIC SRL.

Chapter 5

The Preclinical Phase of Parkinson's Disease

Werner Poewe

The diagnosis of idiopathic Parkinson's disease (IPD) can be made clinically with considerable accuracy when typical motor signs responsive to dopaminergic stimulation are found in the absence of exclusion criteria (Gibb and Lees, 1988). Most patients are in their 60s when this happens, although a diagnosis is established before the age of 40 in approximately 20% of patients. The major source of diagnostic error even in specialized units are parkinsonian syndromes occurring as a presenting feature of neuronal multiple system degenerations, which may share, at least initially, all typical clinical features with IPD. On pathological examination they account for approximately 10% of cases diagnosed as IPD in life (Hughes et al., 1992).

In clinical practice it may be difficult to establish the exact time of clinical manifestation of IPD, since patients may be curiously unaware of early and discrete changes in gait, arm swing, or other signs of motor slowing. It is not uncommon for the accompanying spouse to antedate the occurrence of first signs as described by the patient by 1 to 2 years. Subtle mental changes in terms of slight inertia, loss of energy, and initiative even may have been noticed by those around them even earlier. In the light of increasing efforts to develop neuroprotective treatment strategies for neurodegenerative disorders, early diagnosis becomes a crucial issue. In the case of Parkinson's disease this is all the more critical, as there is evidence to suggest a preclinical phase that possibly may last for decades.

ADVANCES IN RESEARCH ON NEURODEGENERATION, I
D.B. Calne et al.

Evidence for a Prelinical Phase of Parkinson's Disease

Several lines of evidence suggest the existence of a preclinical phase IPD (see Table 1). Soon after the discovery of striatal dopamine loss as the central neurochemical abnormality of IPD (Ehringer and Hornykiewicz, 1961) it became clear that the concentration of dopamine in the putamen must decrease by about 80% for clinical signs to be clearly present (Bernheimer et al., 1973). The corresponding neuronal fallout of pigmented cells in the substantia nigra, pars compacta, must be in the order of 50% to 60% (Fearnley and Lees, 1991). Both results imply a presymptomatic stage of IPD; this is underlined further by additional neuropathological evidence. Some 10% to 15% of those aged above 70 years display nigral zona compacta cell attrition with intraneuronal Lewy bodies in the absence of clinical signs of extrapyramidal dysfunction—so-called incidental Lewy bodies (Forno, 1966; Gibb and Lees, 1988; Fearnley and Lees, 1991). Such cases are likely to represent preclinical Parkinson's disease on the basis of several arguments:

1. Lewy body degeneration in these cases is morphologically identical to that seen in IPD.
2. The degree of nigral neuronal attrition in incidental Lewy body cases is intermediate between normal and IPD brains.
3. The subregional pattern of nigral cell loss is identical to that seen in IPD and different from age-related neuronal fall-out.

Table 1. Evidence for a Preclinical Phase of Parkinson's Disease

1. 80% threshold of striatal DA loss
2. 50% threshold of nigral cell loss
3. Incidental nigral Lewy body degeneration
4. Transient premorbid "unmasking" of symptoms
5. Discordant period of >20 years in concordant MZ twins
6. Preclinical abnormalities on ^{18}F-dopa PET scanning

If incidental Lewy body brains represent preclinical cases, IPD might be 10 to 15 times more prevalent than epidemiological figures based on clinical disease suggest; that is, some 4% of the population aged between 40 and 60 years and 10% to 15% of those aged above 70 years would harbor the disease and should receive neuroprotective therapy once this becomes possible.

Some pieces of clinical observation also point toward the existence of a latent phase of the disease. They include scattered reports of transient "unmasking" of typical signs many years before the onset of IPD that can be obtained from individual observant patients by everyone seeing many cases over the years. Usually this occurs in the context of emotional stress or trauma and consists of transient unilateral limb tremor affecting the side later to become involved in typical IPD (Gowers, 1886; Lees, 1991; see below).

More recently it has become evident that the discordant phase in initially clinically unaffected co-twins of IPD patients can last for more than 20 years, before clinical concordance finally becomes apparent (Duvoisin, 1991). Again, a long latent phase of the illness is suggested for these co-twins. In addition, an ongoing ^{18}F-dopa PET study of clinically unaffected MZ and DZ co-twins of IPD patients so far revealed abnormal putaminal ^{18}F-dopa uptake as a sign of preclinical disease in about one-third of cases (Burn et al., 1991).

The Latent Phase of IPD: 5 or 30 Years?

If there is indeed a presymptomatic phase of IPD, as becomes increasingly likely, how long does it last and how early would protective therapy have to begin? At present there is no clear answer to this. Some of the clinical evidence cited above would point to a latent phase possibly lasting for decades. A preclinical stage of approximately 30 years' duration was suggested on the basis of a comparison of age-specific prevalence rates for incidental Lewy body and clinical Parkinson's disease (Gibb and Lees, 1988). If, according to a hypothesis originally put forward by Calne and Langston (1983), IPD is the result of a biphasic process starting with a destructive nigral insult insufficient in itself to cause overt disease that would only develop as a result of subsequent "normal" age-related neuronal attrition, the rate of the latter could be taken as an index for the possible duration of a latent disease phase. The rate of nigral neuronal attrition with aging has been found to be linear and is between 4% and 6% per decade (Hirai, 1968; McGeer et al., 1977; Fearnley and Lees, 1991). Given the symptomatic threshold of about 50% nigral cell loss and variable degrees of nigral destruction by a hypothetical etiological acute insult, long presymptomatic phases again would seem plausible.

Contrary to this, recent findings by Fearnley and Lees (1991) indicate an exponential rate of nigral neuronal loss in IPD with more than

40% of pigmented neurons disappearing in the first decade of the illness and thus are more compatible with an acute monophasic disease process. Based on their figures of disease-specific neuronal attrition, these authors calculated a preclinical phase of about 5 years' duration. This is considerably shorter than previously discussed and in the range of what the patients' environment often may describe as the time period during which subtle abnormalities not yet appreciated by the patient were actually noticeable. It will be of great significance to observe the latencies between detection of abnormal ^{18}F-dopa PET scans in asymptomatic co-twins of IPD patients and the first occurrence of clear clinical signs.

Are There "Preclinical" Signs of IPD?

Neurobehavioral Antecedents or "Premorbid Personality"

It is now well recognized that subtle and distinctive neuropsychological abnormalities can be detected in newly diagnosed patients with Parkinson's disease of recent onset (Lees and Smith, 1986; Brown and Marsden, 1990). They include difficulties in generating self-paced behavioral responses, particularly problems with the initiation, execution, and completion of cognitive sets as seen in the frequently subnormal performance on tests like the Wisconsin Card Sorting Test (Lees, 1989). Generally, patients seem to do worse than controls whenever the test performance has to be based on internal rather than external cues (Brown and Marsden, 1990). Collectively, this has been interpreted as a sign of compromised frontal lobe function, possibly because of disruption of frontocaudate neuronal circuits (Alexander et al., 1990). Whether or not cognitive slowing or "bradyphrenia" is another early and specific neurobehavioral abnormality in IPD dependent on similar functional systems has remained controversial (Rogers et al., 1987; Poewe et al., 1991), and it remains unclear whether such early cognitive changes are actually present before the onset of the typical motor syndrome.

Episodes of depression occur in up to 50% of IPD patients in the course of their illness, which may be twice the indicence found in other chronically disabling illnesses (Lees, 1989). Some 20% go through periods of major depression before the onset of motor symptoms and depression may antedate clinical diagnosis by more than 20 years (Santamaria et al., 1986).

The most controversial issue in the discussion of neurobehavioral abnormalities in IPD concerns the concept of the presence of distinctive premorbid personality traits in such patients. Ever since the beginning of this century neurologists and psychiatrists have reported on such features and the premorbid parkinsonians were found to be introverted, inflexible, overcontrolled individuals suppressing aggressive tendencies (see Table 2). Although challenged by a large psychological study of more than 100 patients (Riklan et al., 1959), the premorbid personality concept has gained some recent support from studies examining retrospectively the personality features of patients and controls using personality inventories and standardized interviews. Again there was evidence for the patients to have been more controlled, inflexible, introverted, and rigid in thought (Poewe et al., 1983, 1989; Eatough et al., 1990). Some support for the notion of discrete premorbid behavioral abnormalities also emerged from the North American twin study of IPD, where the affected sibling had appeared less dominant and more self-controlled than his unaffected co-twin from early childhood; by the age of 16, he also appeared more nervous, whereas within 10 years prior to the onset of IPD, he had become less aggressive, confident, and light-hearted (Ward et al., 1984).

Although difficult to approach prospectively and despite similar personality findings in other disorders including diabetes, cancer, ulcerative colitis, and essential tremor (Cassileth et al., 1984; Poewe et al., 1990), the possibility that certain behavioral traits could correspond to early and preclinical brain dysfunction of IPD is not entirely unlikely. In particular, premorbid depressive-introverted and obsessional-rigid features might well be linked to preclinical disease. Depression is a well known antecedant of clinical IPD (see above) and has been linked to mesocorticolimbic dopaminergic or brain stem noradrenergic dysfunction and obsessive-compulsive behavior can accompany structural basal ganglia lesions (Laplane et al., 1989).

Transient "Unmasking" of Signs by Emotional Trauma

There are many reports in the literature where first signs of Parkinson's disease seemed to present in the context of emotional stress or trauma. Purves-Stewart in 1898 found that 6 of his 28 patients gave a history of fright or emotional upset as the cause of their disease; one man triggered his first notice of stiffness in his left leg the day after having gone through the terror of a lion escaping from its cage at a zoo he was visiting. Similar

Table 2. Premorbid Personality in Parkinson's Disease

Camp (1913): "Industriousness," "moralistic"
Cohen-Booth (1935): "Suppression of aggressive tendencies"
Sands (1942): "Masked personality"
Mitscherlich (1960): "Neurotic-compulsive character"
Korten and Kettering (1972): "Reliable, sense of duty, moral rigidity"
Mouren et al. (1983): "Meticulousness and perfectionism"
Poewe et al. (1983): "Overcontrolled, introverted, depressed"
Ward et al. (1984): "Self-controlled, quiet, unhappy"

observations of rest tremor first appearing in the context of emotional shock are also mentioned by other early authors (Gowers, 1886; Lewy, 1923) and it is not uncommon for patients today to cite similar events as possible triggering factors for their illness. Given a natural tendency to remember emotionally upsetting events, such dating of first symptoms by the patients might well be an artefact. More important for the concept of latent PD, however, are reports by observant patients about the transient appearance of typical signs of PD many years before the proper start of the illness. The following observation related by one of my patients may serve to illustrate this point:

This 56-year-old gentlemen was first referred for evaluation of his drug regimen of levodopa plus amantadine 3 years after typical levodopa-responsive IPD has begun with resting tremor of his right hand. By the time I saw him all cardinal signs of IPD were present, with the right-sided extremities being more severely affected. He clearly remembered an episode of his youth, when at about the age of 16 he had been caught in a thunderstorm on his way home in a village in the Tyrolean Alps. He had sought shelter under a tree but this was soon hit by lightning, causing a major branch to break off and crash down, only narrowly missing the young man. He had fled the place in horror; upon reaching his home he first noticed involuntary shaking of his right arm. He was certain that it had affected only the right side and had been present more or less continuously for some days before completely disappearing again. As far as he could remember it had been the same kind of shaking that decades later ushered in his IPD.

Many clinicians seeing patients with IPD are able to relate similar anecdotes, and there is more evidence pointing toward a possible relation between emotional trauma and manifestation of parkinsonian signs. Lees

(1991) has repeatedly drawn attention to Grinker and Spiegel's report (1943) on neurological problems of American soldiers during the Tunisian campaign where cases of shell-shock are described as assuming a stooped simian posture with mask-like expressionless faces, occasionally dripping saliva from a mouth held open, and showing coarse limb tremors. It is interesting to note that Parkinson's disease was found to be about 5 times as prevalent than to be expected by epidemiological data in a population of 4684 British ex-prisoners of war from the Far East (Gibberd and Simmonds, 1980).

Subtle Sensorimotor Signs Antedating Clinical Diagnosis

Observant patients sometimes report sensory alterations of the arm or leg later to be first affected by IPD that antedate clearly recognizable manifestations of the disease by several years. These may include pulling sensations in muscles, feelings of "inner" trembling of a limb, vague dysesthesias, and pain. Alternatively, those around the patient may notice subtle changes of gait, stride, or arm swing before the sufferer himself has taken note of them. Lees (1991) has analyzed video material from TV footage covering football matches played by Ray Kennedy, a famous soccer player now afflicted with IPD. Slight motor changes while playing were clearly apparent in this patient some 5 years before the clinical diagnosis could be confidently established. Horstink and colleagues (Gonera et al., 1991) have attempted a systematic study of such "preclinical" sensorimotor syndromes in IPD patients by checking family doctors' notes covering the 10-year period prior to clinical diagnosis and comparing frequency of and reasons for doctor's visits to those of age-matched nonneurological controls from the same doctor's registry during the same time period. They found evidence for an increased incidence of mood changes and musculoskeletal pain in the patient group (see Table 3), but unfortunately their study lacked precise information on the exact time relation of such complaints to the onset of clinical IPD signs.

Table 3. Symptoms Preceding the Clinical Diagnosis of Parkinson's Disease

Mood disorder	59 vs. 18
Myalgia	44 vs. 17
Shoulder pain	18 vs. 7
Other sensory complaints	11 vs. 2

Years −10 to 0; 62 patients vs. 56 controls; Gonera et al. (1991).

Taking together the information from such studies and anecdotical reports by patients and relatives, it would seem probable that subtle motor changes heralding the onset of IPD might be detected around 5 years before classical onset of the disease. As a consequence, it would seem feasible to antedate a clinical diagnosis of IPD by such a time factor compared to current practice. Developing an appropriate clinical test battery for early signs should lead to the definition of "at-risk" persons, for whom further diagnostic markers would be needed, before they might receive protective drug treatment.

Diagnostic Markers?

Currently no diagnostic markers for IPD exist. Nevertheless, some test findings may substantiate a presumptive diagnosis of IPD in a person exhibiting clinical signs. These include reduced putaminal ^{18}F-dopa uptake on PET scanning and a positive response to dopaminergic stimulation (apomorphine or L-dopa test). While dopaminergic responsiveness may be impossible to assess reliably in a patient exhibiting vague sensorimotor signs (possible preclinical IPD), ^{18}F-dopa PET scans could yield positive results in such instances. Calne and co-workers (1986) were first to demonstrate subclinical abnormalities of ^{18}F-dopa PET scans in four individuals exposed to the nigral neurotoxin MPTP and the Hammersmith Hospital PET group have studied 11 asymptomatic co-twins of IPD probands and found clearly abnormal putaminal ^{18}F-dopa uptake in four cases. Sawle et al. (1991) have calculated cut-off values for striatal ^{18}F-dopa uptake that would allow for distinction between pathological and normal low values. However, PET technology is of restricted availability and a research rather than a diagnostic screening tool. First advances toward studying the functional status of the nigrostriatal dopamine system by more widely applicable SPECT techniques have been made with the use of D_2-receptor binding ligands, but currently there are no data to suggest that this might gain significance in the detection of preclinical abnormalities specific for IPD. Chemical markers would form a more accessible diagnostic aid for screening possible at-risk persons, but none of those abnormalities found in IPD patients have so far proven specific. With more families with autosomal dominant Lewy body Parkinson's disease (Golbe et al., 1990), reported linkage studies might eventually lead to genetic markers at least for subgroups of IPD.

How to Define an "At-Risk" Population for IPD

Advances in the definition of diagnostic markers with some specifity for IPD and the future development of neuroprotective treatment strategies require criteria for the clinical identification of at-risk populations to be screened and possibly treated.

Currently there is a lack of such clinical criteria. Table 4 lists the mosaic of components from which a definition of an at-risk group for IPD might eventually arise. Further studies must investigate sets of clinical tests of subtle motor deficits in the earliest stages of IPD and those found to be most discriminant might then be applied to putative risk groups (e.g., first degree relatives of IPD patients). Abnormal scorers should subsequently be followed up and possibly also undergo ^{18}F-dopa PET studies. One goal that might be reached by such research protocols could be the eventual antedating of a clinical IPD diagnosis by several years compared to current mean ages at diagnosis of IPD patients.

Table 4. Toward a "Preclinical" Diagnosis of Parkinson's Disease

Define "at-risk" population
- first-degree relatives of PD patients
- nonsmoker
- anancastic personality disorder
- depression
- vague sensorimotor symptoms
- isolated tremor
- N.N.

Apply diagnostic screening test
- ^{18}F-dopa PET
- genetic marker
- chemical marker
- N.N.

Antedate diagnosis and treatment by 5 years

References

Alexander G, DeLong M, Stück P (1986): Parallel organisation of functionally segregated circuits linking basal ganglia and cortex. *Annu Rev Neurosci* 9:357–381.

Bernheimer B, Birkmayer W, Hornykiewicz O, et al. (1973): Brain dopamine and syndromes of Parkinson and Huntington. Clinical, morphological and neurochemical correlations. *J Neurol Sci* 20:415–455.

Burn DJ, Mark MH, Playford ED, Maraganore J, Zimmermann TR, Duvoisin RC, Harding AE, Marsden CD, Brooks DJ (1991): Parkinson's disease in twins: an 18F-DOPA-study (Abstr.). *Proc 10th Internat Symp Parkinson's disease.* Tokyo.

Calne DB, Langston JW (1983): Aetiology of Parkinson's disease. *Lancet* ii:1457–1459.

Cassileth BR, Lusk EJ, Strouse TB, et al. (1984): Psychological status in chronic illness. *N Eng J Med* 311:506–511.

Duvoisin RC (1991): The genetics of Parkinson's disease: a review. In: *International Workshop Berlin Parkinson's Disease.* Rinne UK, Nagatsu T, Horowski R, eds. Bussum: Medicom Europe.

Eatough VM, Kempster PA, Lees AJ, Stern GM (1990): Pre-morbid personality in idiopathic Parkinson's disease. In: *Advances in Neurology*, Vol 53. Streifler et al., eds. New York: Raven Press.

Ehringer H, Hornykiewicz O (1960): Verteilung von Noradrenalin und Dopamin (3-Hydroxytyramin) im Gehirn des Menschen und ihr Verhalten bei Erkrankungen des extrapyramidalen Systems. *Klin Wochenschr* 38:1236–1239.

Farnley JM, Lees AJ (1991): Ageing and Parkinson's disease: substantia nigra rational selectivity. *Brain* 114:2283–2301.

Forno LS (1969): Concentric hyalin intraneuronal inclusions of Lewy type in the brain of elderly persons (50 incidental cases). Relationship to Parkinsonism. *J Am Geriatr Soc* 17:557–575.

Gibb WRG, Lees AJ (1988): The relevance of the Lewy body to the pathogenesis of idiopathic Parkinson's disease. *J Neurol Neurosurg Psychiatry* 51:745–752.

Gibberd FB, Simmonds JP (1980): Neurological disease in ex-far-east prisoners of war. *Lancet* ii:135.

Golbe LI, Cody RA, Duvoisin RC (1986): Smoking and Parkinson's disease—search for a dose-response relationship. *Arch Neurol* 43:774–778.

Golbe LI, DiIorio G, Bonavita V, Miller DC, Duvoisin RC (1990): Autosomal dominant Parkinson's disease. *Ann Neurol* 27:276–282.

Gonera EG, Horstink MWIM, Berger HJC (1991): The existence and the symptoms of preclinical PD (Abstr.) *Proc. 10th International Symposium on Parkinson's Disease*, Tokyo.

Gowers WR (1886): *A Manual of Diseases of the Nervous System.* London: J & A Churchill.

Grinker RR, Spiegel JP (1943): War neuroses in North Africa: the Tunisian campaign. Prepared and distributed for the Air-surgeon Army Air Forces by the Joseph Mocy Jr Foundation, 14–17, 40–44.

Hirai S (1968): Histochemical study on the regressive degeneration of the senile brain, with special reference to the aging of the substantia nigra. *Adv Neurol Sci* 12:845–849.

Hughes AJ, Daniel SE, Kilford L, Lees AJ (1991): The accuracy of the clinical diagnosis of Parkinson's disease: a clinico-pathological study of 100 cases. *J Neurol Neurosurg Psychiatry* 55:181–184.

Laplane D, Levasseur M, Pillon V, et al. (1989): Obsessive-compulsive and other behavioural changes with bilateral basal ganglia lesions. *Brain* 112:649–725.

Lees AJ (1989): The neurobehavioural abnormalities in Parkinson's disease and their relationship to psychomotor retardation and obsessional compulsive disorders. *Behav Neurol* 2:1–4.

Lees AJ (1991): The early detection of Parkinson's disease. In: *International Workshop Berlin Parkinson's Disease*, Rinne UK, Nagatsu T, Horowski R, eds. Bussum: Medicom Europe, pp. 156–170.

Lees AJ, Smith E (1983): Cognitive deficits in the early stages of Parkinson's disease. *Brain* 106:257–270.

Lewy FH (1923): *Die Lehre vom Tonus und der Bewegung*. Berlin: Springer.

McGeer PL, McGeer EG, Suzuki JS (1977): Aging and extrapyramidal function. *Arch Neurol* 34:33–35.

Poewe W, Berger W, Benke Th, Schelosky L (1991): High-speed-memory scanning in Parkinson's disease: adverse effects of levodopa. *Ann Neurol* 29:670–673.

Poewe W, Karamat E, Kemmler GW, Gerstenbrand F (1990): The premorbid personality of patients with Parkinson's disease: a comparative study with healthy controls and patients with essential tremor. *Adv Neurol* 53:339–342.

Poewe WG, Gerstenbrand F, Karamat E, Schmidhuber-Eiler B (1989): The premorbid personality in patients with Parkinson's disease. An early sign of their disease. In: *Key Topics in Brain Research. Early Diagnosis and Preventive Therapy in Parkinson's Disease*, Przuntek H, Riederer P, eds. Vienna: Springer.

Poewe WH, Gerstenbrand F, Ransmayr G, Plorer S (1983): Pre-morbid personality of Parkinson's patients. *J Neural Transm*, suppl. 19:215–224.

Purves S (1898): Paralysis agitans; with an account of a new symptom. *Lancet* 1258–1260.

Riklan M, Weiner H, Diller L (1959): Somato-psychologic studies in Parkinson's disease. 1. An investigation into the relationship of certain disease factors in psychological functions. *J Nerv Ment Dis* 129:263–267.

Rogers D, Lees AJ, Smith E, Trimble M, Stern GM (1987): Bradyphrenia in Parkinson's disease and psychomotor retardation in depressive illness. *Brain* 110:761–776.

Santamaria J, Tolosa E, Valles A (1986): Parkinson's disease with depression: a possible subgroup of idiopathic parkinsonism. *Neurology* 36:1130–1133.

Sawle GV, Playford ED, Burn DJ, Brookds DJ, Frackowiak RSJ (1991): Can pet identify preclinical Parkinson's disease amongst a population of healthy volunteers? (Abstr.) *Proc. 10th International Parkinson's Disease*, Tokyo.

Ward CD, Duvoisin RC, Ince SE (1984): Parkinson's disease in twins. *Adv Neurol* 40:341–344.

Chapter 6

Mitochondria in Parkinson's Disease

Yoshikuni Mizuno, Shin-ichirou Ikebe,
Nobutaka Hattori, and Hiroyo Yoshino

Recently, mitochondria have become one of the hottest targets for the research on Parkinson's disease. This interest originated from the observation on MPTP (1-methyl-4-phenyl-1,2,3,6-tetrahydropyridine)-induced parkinsonism. In this chapter we briefly review the mechanism of mitochondrial failure in MPTP-induced parkinsonism, and recent research on mitochondria of Parkinson's disease. We also briefly discuss the mechanisms of nigral death in Parkinson's disease.

Mechanism of Mitochondrial Failure in MPTP-Induced Parkinsonism

MPTP is an unique substance in that systemically administered MPTP causes selective damage of substantia nigra (see Figure 1). This selectivity depends on the active uptake of its oxidative metabolite, MPP^+ (1-methyl-4-phenylpyridinium ion), into dopaminergic neurons, the oxidation being catalyzed by monoamine oxidase B in astrocytes (Chiba et al., 1984a). This active uptake takes place against the concentration gradient of MPP^+ utilizing the energy of ATP through the dopamine uptake site (Chiba et al., 1984b). Thus, MPP^+ is highly concentrated within nigrostriatal dopaminergic neurons (Irwin and Langston, 1985). Non-dopaminergic neurons do not have an active uptake site for dopamine neither for MPP^+. This is the basis for the selective degeneration of dopaminergic cells in this model. Within dopaminergic neurons, MPP^+ is further concentrated into mitochondria by active transport (Ramsay and Singer, 1986). The mechanism of this active transport is understood as

ADVANCES IN RESEARCH ON NEURODEGENERATION, I
D.B. Calne et al.

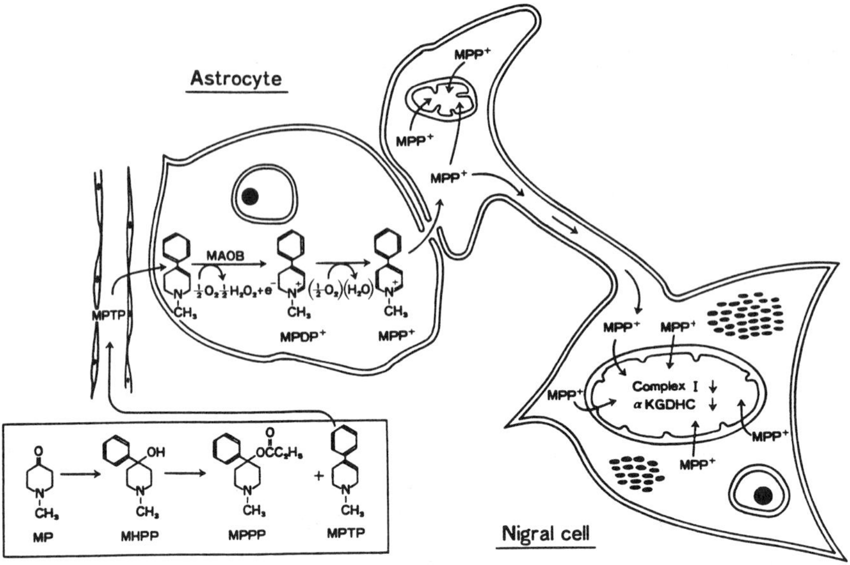

Figure 1 Schematic representation of the mechanism of selective accumulation of MPP$^+$ within nigral mitochondria.

follows: As electron transport takes place along the electron transport complexes, protons are expelled out of the inner membrane, producing electrochemical potential difference between the inside and outside of the inner membrane with the inside negatively charged. According to this electrochemical gradient, MPP$^+$ is transported into mitochondria until electrochemical equilibrium is reached. Not all the positively charged substances are actively transported into mitochondria. Appropriate lipid solubility seems to be an additional prerequisite.

Within mitochondria, MPP$^+$ inhibits mitochondrial respiration by inhibiting complex I of the electron transport system (Mizuno et al., 1986, 1987a) and the α-ketoglutarate dehydrogenase complex of the TCA cycle (Mizuno et al., 1987b). In this experiment, mitochondria isolated from whole brains of B57/BL mice were used as an enzyme source after freeze-thawing of mitochondria. Freeze-thawing is necessary to measure the full activity, because it modifies the inner membrane of mitochondria so as for NADH, one of the substrates for complex I, to penetrate the inner membrane easily. MPTP was a more potent inhibitor of complex I than MPP$^+$, in that the IC_{50} of MPTP was 0.76 mM in contrast to 3.4 mM for MPP$^+$. Even with the higher IC_{50}, MPP$^+$ was more toxic than MPTP to intact mitochondria. Actually, MPTP did not inhibit mitochondrial respiration with the concentration that can substantially inhibit complex I of freeze-thawed mitochondria. MPTP does not seem to cross the inner mitochondrial membrane sufficiently enough to inhibit mitochondrial respiration. Also, some neurotoxins (mainly their reduced forms) do not penetrate the inner membrane well. On the other hand, 0.05 mM of MPP^+ was able to inhibit the mitochondrial state 3 respiration to approximately 10% and complex I activity to 50% of respective controls when MPP^+ was preincubated with intact mitochondria together with substrates of the TCA cycle such as glutamate and ADP (Mizuno et al., 1988). As 3.4 mM of MPP^+ was necessary to obtain 50% inhibition of complex I activity of freeze-thawed mitochondria, the concentration of MPP^+ inside of intact mitochondria preincubated with 0.05 mM of MPP^+ must be close to that concentration, indicating approximately 60-fold concentration of MPP^+ within mitochondria. Actually, Ramsay et al. (1987) estimated the maximum difference in the concentration of MPP^+ between nigral mitochondria and the tissues surrounding substantia nigra to reach 800-fold. This unique property of MPP^+ should always be kept in mind when a candidate toxin is sought for Parkinson's disease.

Mitochondrial Abnormalities in Parkinson's Disease

Table 1 summarizes recent reports on mitochondria of Parkinson's disease. Schapira et al. (1989) reported a loss of complex I activity in Parkinson's disease for the first time. We were working on the same subject, but in our hand, the difference in the activity of complex I between Parkinson's disease and the control did not reach the statistical significance (Mizuno et al., 1990). This may in part be due to the difference in the method and the tissues used. We analyzed striatal mitochondria. Schapira et al. used total homogenates of substantia nigra. Another factor that should be considered is the post-mortem instability of complex I. The activity of complex I decreases as the post-mortem delay before freezing the brain increases. In addition, hypotension and hypoxia in the agonal stage probably will influence the activity significantly. Evaluation of these factors is practically impossible because the condition of the agonal state is different from case to case. Therefore, even matched for the age, sex, and post-mortem delay, there remains a problem in the assay of the enzymatic activity of complex I.

Table 1. Mitochondrial Abnormalities Reported in the Literature

1989 Schapira et al., Decrease in complex I activity
1989 Bindoff et al., Complex I, II, IV deficiencies in skeletal muscles
1989 Mizuno et al., Loss of complex I subunits (Western blotting)
1989 Parker et al., Decrease in complex I activity in platelets
1990 Schapira et al., Decrease in complex I activity, normal mtDNA
1990 Mizuno et al., Decrease in Complex III activity
1990 Schapira et al., Normal mtDNA (Southern blotting)
1990 Ikebe et al., Increase in deleted mtDNA (PCR)
1990 Lestienne et al., Normal mtDNA (Southern blotting)
1990 Schapira et al., Disease specific complex I deficiency
1990 Ozawa et al., Quantitative determination of deleted mtDNA
1990 Ozawa et al., Clustering of point mutations in mtDNA
1991 Schoffner et al., Oxidative phosphorylation defect in muscle
1991 Hattori et al., Immunohistochemistry of complex I

Therefore, we analyzed amount of subunits composing complex I by Western blot analysis and immunohistochemistry. Complex I is composed of 25 subunits. With the Western blotting, we found decreases in the subunits of complex I with the molecular weight between 24 and 30 kD in four out of five patients with Parkinson's disease (Mizuno et al., 1989). Then, we studied substantia nigra of eight patients and seven control subjects in detail using antisera against complexes I, II, III, and IV by immunohistochemistry (Hattori et al., 1991). Practically all the nigral cells were stained well for complexes III and IV in both parkinsonian and control subjects. When stained for complex II, small proportions of nigral neurons showed reduced immunostaining in three out of eight patients with Parkinson's disease. When stained for complex I, neurons with reduced immunostaining were seen not only in parkinsonian patients, but also in all the control subjects studied. Actually, nigral melanin–containing cells were a mixture of intensely stained to barely or poorly stained neurons. Therefore, we made a semiquantitative analysis. All the melanin–containing nigral neurons in one of the slices of the midbrain were classified arbitrarily into three groups according to the intensity of immunostaining for complex I; that is, intensely stained neurons (normal), weakly stained neurons, and barely stained ones (including those with no apparent immunostaining). The proportion of weakly plus barely stained neurons among total nigral melanin–containing neurons was significantly higher in Parkinson's disease (35.6 ± 17.6%) than control subjects (13.6%

± 4.2%). These findings suggest the presence of selective decrease in complex I in Parkinson's disease. The presence of similar decrease in lesser amount in control patients may be one of the reasons for age-related decline in the nigral cells.

As 13 of the electron transport complex subunits (seven subunits of complex I, apocytochrome b, cytochrome c oxidase subunits I, II, and III, and ATPase subunits 6 and 8) are coded by mitochondrial DNA (mtDNA), we analyzed mtDNA in patients with Parkinson's disease (Ikebe et al., 1990). Total DNA was extracted from striata, and fragments of mtDNA were amplified by the PCR method. The amplified fragments were separated by electrophoresis and were detected fluorographically after staining with ethidium bromide. When two synthetic oligonucleotide primers that encompass the ND5 and ATPase 8 genes were used, the normal 5.75-kb fragment was amplified from all the subjects studied. In addition, a shorter fragment of 0.77 kb was amplified from mutant mtDNA in all the parkinsonian patients (n=5) and in two aged control subjects out of six controls. The amount of deleted mtDNA was estimated to be approximately 5% of total striatal mtDNA (Ozawa et al., 1990). By the direct sequencing method, this deletion was found to be present between the 13-bp directly repeated sequence in the ND5 and ATPase 8 genes (Ikebe et al., 1990). This area represents a hot spot of mutations among many patients with mitochondrial myopathies (Tanaka et al., 1989). Our preliminary sequencing of parkinsonian mtDNA revealed several nonsynonymous point mutations that were not found in control subjects (Ozawa et al., 1991). However, we could not detect a point mutation common to Parkinson's disease.

Mechanism of Mitochondrial Abnormalities in Parkinson's Disease

The cause of Parkinson's disease may be a localized mitochondrial respiratory failure secondary to the long-term accumulation of noxious stimuli. Noxious stimuli may be toxic substances, accelerated aging, and other unknown factors. Mixtures of well stained to poorly stained nigral neurons for complex I in Parkinson's disease suggest this hypothesis. When the mitochondrial electron transport is slowed down by such toxins, electrons will accumulate in higher amount somewhere along the complex I. Those excessive electrons will reduce ferric iron in the iron-sulfur cluster to ferrous iron, which may stimulate the production of cytotoxic free radicals such as hydroxyl radicals (·OH) according to the following equation;

$$Fe^{++} + H_2O_2 \longrightarrow Fe^{+++} + \cdot OH + OH^- \quad \text{(the Fenton reaction).}$$

Hydroxyl radicals can oxidize nonspecifically numbers of molecules causing dysfunctions of those substances. In addition, hydroxyl radicals can act on deoxyguanosine molecules of mtDNA to form 8-hydroxy-deoxyguanosine predisposing mtDNA to G·C to T·A transversion mutation (Hayakawa et al., 1991).

mtDNA is small in size without protein coat or repair mechanism. This may be one of the reasons for the high frequency of gene mutation in mtDNA. mtDNA seems more vulnerable to mutagenic damage from exogenous toxins or age-related insults, including free radicals. We found mutations and a deletion in mtDNA and loss of complex I subunits in parkinsonian patients, which supports the hypothesis that progressive accumulation of noxious stimuli in nigral mitochondria eventually may damage substantia nigra to cause Parkinson's disease. mtDNA mutations may lead to the instability of DNA and eventual deletions, such as in yeast mtDNA (Linnane et al., 1989). In addition, mtDNA point mutations and deletions will derange the assembly of electron transport complexes, resulting in further disturbance of the mitochondrial respiration. Thus, a vicious cycle may be established that eventually may lead to ATP depletion. These processes are schematically represented in Figure 2.

Consequence of ATP Depletion

There are numbers of reactions in which ATP plays an important role. Without ATP, cells cannot survive. Among many reactions, perhaps one of the most important enzymes that directly regulates the cellular milieu for survival is Na-K ATPase. This enzyme is essential for the maintenance of high intracellular K^+ and low Na^+ concentrations. Na^+ is actively exported out of cells exchanging with K^+ using the energy of ATP. This is catalyzed by Na-K ATPase. As the activity of Na-K ATPase decreases secondary to ATP depletion, the maintenance of low intracellular Na^+ concentration becomes difficult. Then, Na^+ must be expelled out of cells exchanging with Ca^+ through a calcium channel. Increase in Ca^+ influx will induce many deleterious reactions, including the stimulation of calcium-activated proteinase such as calpaine. This may be the

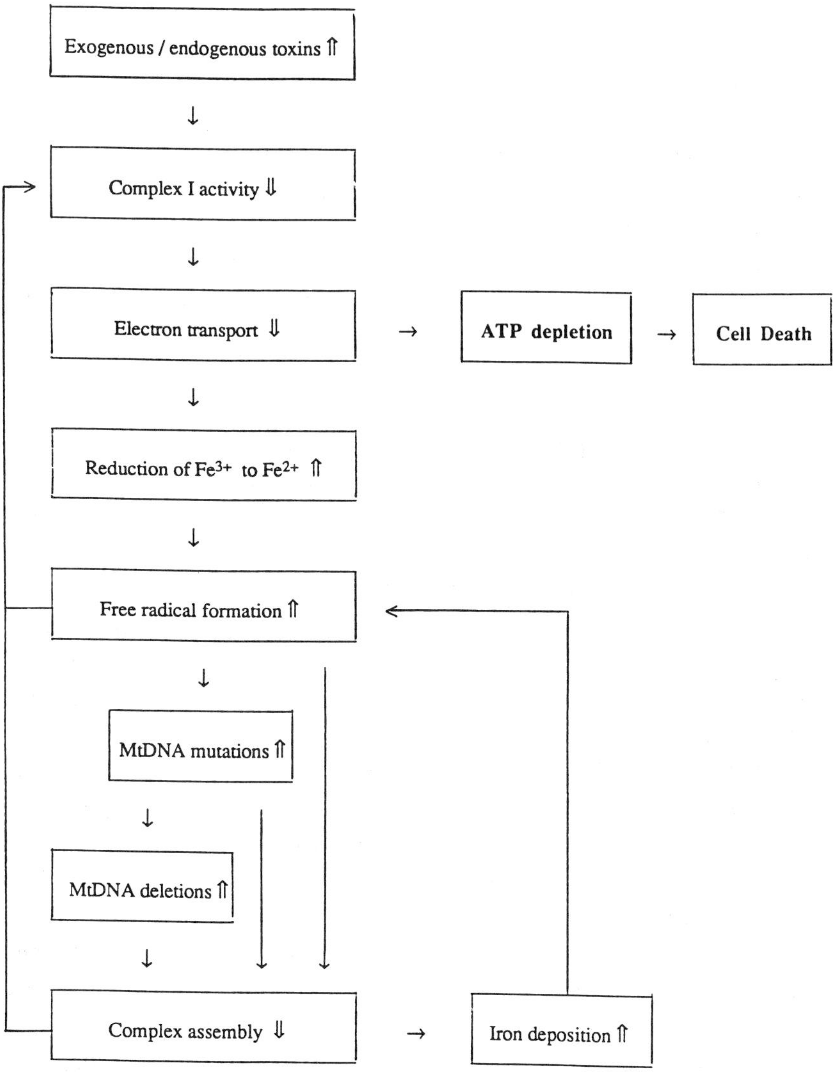

Figure 2 Schematic representation of nigral cell death in Parkinson's disease, a hypothesis.

initial step of cellular degradation resulting from mitochondrial respiratory failure.

Candidate Toxins for Parkinson's Disease

It is amazing that an exogenous substance such as MPTP has a route to be concentrated selectively in the nigral mitochondria. Suppose similar toxins are present in our environment; they may well be accumulated into nigral cells causing premature degeneration of substantia nigra in predisposed persons. For any candidate toxins, properties to be oxidized by monoamine oxidase B, to be actively taken up into dopaminergic cells, and to be actively transported into mitochondria seem necessary for them to cause nigral death in the similar way as MPP^+. Numbers of direct derivatives or analogues of MPTP were found to be toxic to mitochondria (Sayre et al., 1990); however, none of them have been shown to be accumulated in parkinsonian brains. Among the substances found in brain with a structural similarity to MPTP, tetrahydroisoquinolines (Figure 3) and beta-carbolines have been studied most extensively. Some of them are toxic to mitochondrial respiration, thus inhibiting the activity of complex I (Suzuki et al., 1990). Tetrahydroisoquinolines (TIQs) are present in many plants and foods. They can easily cross the blood-brain barrier, and are found in the brain (Niwa et al., 1987). Although TIQ itself does not seem to be very toxic, TIQ can be N-methylated by an N-methyl-transferase (Naoi et al., 1989). N-methyl-TIQ (NMTIQ) can be oxidized by monoamine oxidase A and B, like MPTP, to yield N-methyl-tetrahydroquinolinium ion ($NMIQ^+$) (Naoi et al., 1989). We tested toxicity to NMTIQ and $NMIQ^+$ against mitochondria. Results are summarized in Table 2. Similar to MPTP, NMTIQ did not inhibit state 3 respiration (of intact mitochondria), but it inhibited complex I activity of freeze-thawed mitochondria. On the other hand, $NMIQ^+$ inhibited both state 3 respiration (of intact mitochondria) and complex I activity of freeze-thawed mitochondria to approximately the same extent. Therefore, $NMIQ^+$ appears to be toxic to mitochondrial respiration, but NMTIQ does not. This relationship is similar to that of MPTP and MPP^+. The difference is the fact that the toxicity of MPP^+ against state 3 respiration is far greater than that against complex I activity of freeze-thawed mitochondria. No such difference was noted in the toxicity of $NMIQ^+$, indicating that $NMIQ^+$ is not so much concentrated within mitochondria as MPP^+. Therefore, it does not seem likely that $NMIQ^+$ alone can

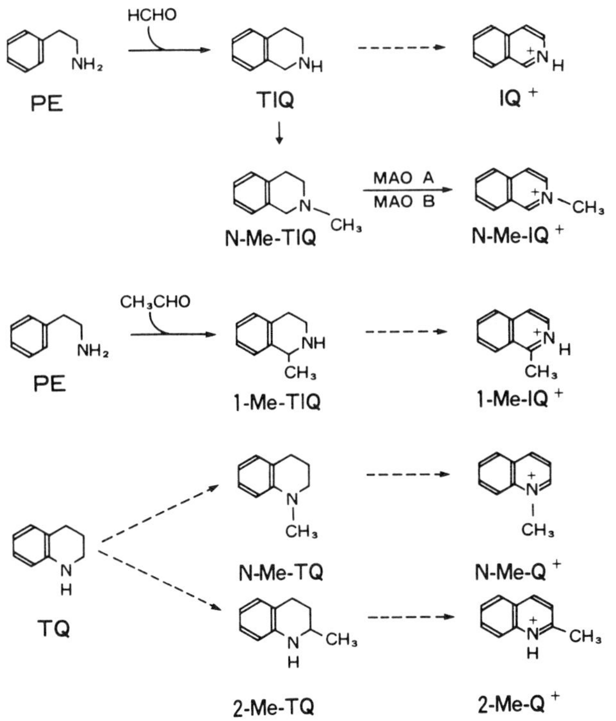

Figure 3 Structures and the relationship of tetrahydroisoquinolines. Reprinted with permission of Excerpta Medica, Ltd. from Mizuno Y (1990): Neurotoxins and Parkinson's disease. In: *Senile Neurodegeneration and Neurotransmitters*, Nagatsu T, ed. Tokyo: Excerpta Medica, pp. 51–64.

induce premature nigral cell death. But if several such compounds are acting together, they may be able to kill the nigral cells.

Table 2. Effects of *N*-Methyl-Tetrahydroisoquinoline and *N*-Methyl-Tetrahydroisoquinolinium Ion on Mitochondria

	State 3 Respiration	Complex I Activity
Control	67.8[a]	172[b]
N-methyl-TIQ (1 mM)	111.6%	75.1%
N-methyl-IQ$^+$ (1 mM)	24.8%	38.8%

Expressed in % of the respective controls.

[a] Nanomole O_2/min/mg protein (intact mitochondria).

[b] Nanomole NADH oxidized/min/mg protein (freeze-thawed mitochondria).

Summary

We have reviewed a current concept on the mechanism of nigral cell death in Parkinson's disease. Mitochondrial complex I abnormalities reported in Parkinson's disease include decrease in the enzymatic activity, loss of subunits, increase in deleted mtDNA, and mtDNA point mutations. These findings are consistent with the concept that Parkinson's disease may be caused by a progressive accumulation of exogenous and endogenous toxins within nigral mitochondria resulting in energy crisis. Further studies along this line would seem to be important in order to elucidate the pathogenesis of Parkinson's disease.

Acknowledgments. Our studies reviewed in this chapter were performed with the collaboration of Drs. T. Ozawa and M. Tanaka (Department of Biomedical Chemistry of Nagoya University), Drs. Y. Kagawa and S. Ohta (Department of Biochemistry, Jichi Medical School), Dr. N. Sone (Kyushu Institute of Technology), Drs. Y. Ohya and S. Takamiya (Department of Parasitology, Juntendo University School of Medicine), Dr. K. Kita (University of Tokyo Institute of Medical Science), Drs. M. Yoshida, K. Suzuki, and T. Saitoh (Department of Neurology, Jichi Medical School), Drs. M. Naoi and W. Maruyama (Nagoya Institute of Technology), and Dr. T. Sato (Department of Neurology, Juntendo University School of Medicine). The authors thank these scientists for this collaboration.

References

Bindoff, LA, Birch-Machin M, Cartlidge NEF, Parker WD, Jr, Turnbull DM (1989): Mitochondrial function in Parkinson's disease. *Lancet* 2:49.

Chiba K, Trevor AJ, Castagnoli N, Jr (1984a): Metabolism of the neurotoxic tertially amine, MPTP, by brain monoamine oxidase. *Biochem Biophys Res Commun* 120:574–578.

Chiba K, Trevor AJ, Castagnoli N, Jr (1984b): Active uptake of MPP^+, metabolite of MPTP, by brain synaptosomes. *Biochem Biophys Res Commun* 128: 1229–1232.

Hattori N, Tanaka M, Ozawa T, Mizuno Y (1991): Immunohistochemical studies on complex I, II, III, and IV of mitochondria in Parkinson's disease. *Ann Neurol* 30:563–571.

Hayakawa M, Ogawa T, Sugiyama S, Tanaka M, Ozawa T (1991): Massive conversion of guanosine to 8-hydroxy-guanosine in mouse liver mitochondrial DNA by administration of azidothymidine. *Biochem Biophys Res Commun* 176:87–93.

Ikebe S, Tanaka M, Ohno K, Sato W, Hattori K, Kondo T, Mizuno Y, Ozawa T (1990): Increase of deleted mitochondrial DNA in the striatum in Parkinson's disease and senescence. *Biochem Biophys Res Commun* 170:1044–1048.

Irwin I, Langston JW (1985): Selective accumulation of MPP^+ in the substantia nigra: a key to neurotoxicity? *Life Sci* 36:207–212.

Linnane AW, Marzuki S, Ozawa T, Tanaka M (1989): Mitochondrial DNA mutations as an important contributor to an ageing and degenerative diseases. *Lancet* 1:642–645.

Mizuno Y, Ohta S, Tanaka M, Takamiya S, Suzuki K, Sato T, Oya H, Ozawa T, Kagawa Y (1989): Deficiencies in Complex I subunits of the respiratory chain in Parkinson's disease. *Biochem Biophys Res Commun* 163:1450–1455.

Mizuno Y, Sone N, Saitoh T (1986): Dopaminergic neurotoxin, MPTP and MPP^+, inhibit mitochondrial NADH-ubiquinone oxidoreductase activity. *Proc Jpn Acad Ser B* 62:261–263.

Mizuno Y, Sone N, Saitoh T (1987a): Effects of 1-methyl-4-phenyl-1,2,3,6-tetrahydropyridine and 1-methyl-4-phenylpyridinium ion on activities of enzymes in the electron transport system in mouse brain. *J Neurochem* 48:1787–1793.

Mizuno Y, Sone N, Saitoh T, (1987b): Inhibition of mitochondrial alpha-ketoglutarate dehydrogenase by 1-methyl-4-phenylpyridinium ion. *Biochem Biophys Res Commun* 143:971–976.

Mizuno Y, Suzuki K, Sone N, Saitoh T (1988): Studies on the toxicity of 1-methyl-4-phenylpyridinium ion (MPP^+) against mitochondria of mouse brain. *J Neurol Sci* 86:97–110.

Mizuno Y, Suzuki K, Ohta S (1990): Postmortem changes in mitochondrial respiratory enzymes in brain and a preliminary observation in Parkinson's disease. *J Neurol Sci* 96:49–57.

Naoi M, Matsuura S, Takahashi T, Nagatsu T (1989): A *N*-methyltransferase in human brain catalyzes *N*-methylation of 1,2,3,4-tetrahydroisoquinoline into *N*-methyl-1,2,3,4-tetrahyroisoquinoline, a precursor of a dopaminergic neurotoxin *N*-methylisoquinolinium ion. *Biochem Biophys Res Commun* 161:1213–1219.

Naoi M, Takahashi T, Parvez H, Kabeya R, Taguchi E, Yamaguchi K, Hirata Y, Minami M, Nagatsu T (1989): *N*-methylisoquinolinium ion as an inhibitor of tyrosine hydroxylase, aromatic *L*-amino acid decarboxylase and monoamine oxidase. *Neurochem Int* 15:315–320.

Niwa T, Takeda N, Kaneda N, Hashizume Y, Nagatsu T (1989): Presence of tetrahydroisoquinoline and 2-methyl-tetrahydroisoquinoline in parkinsonian and normal human brains. *Biochem Biophys Res Commun* 144:1084–1089.

Ozawa T, Tanaka M, Ikebe S, Ohno K, Kondo T, Mizuno Y (1990): Quantitative determination of deleted mitochondrial DNA relative to normal DNA in parkinsonian striatum by a kinetic PCR analysis. *Biochem Biophys Res Commun* 172:483–489.

Ozawa T, Tanaka M, Sugiyama S, Ino H, Ohno K, Hattori K, Ohbayashi T, Ito T, Deguchi H, Kawamura K, Nakane Y, Hashiba K (1991): Patients with idiopathic cardiomyopathy belong to the same mitochondrial DNA gene family of Parkinson's disease and mitochondrial encephalomyopathy. *Biochem Biophys Res Commun* 177:518–525.

Parker WD, Jr, Boyson SJ, Parks JK (1989): Abnormalities of the electron transport chain in idiopathic Parkinson's disease. *Ann Neurol* 26:719–723.

Ramsay RR, McKewn KA, Johnson EA, Booth RG, Singer TP (1987): Inhibition of NADH oxidation by pyridine derivatives. *Biochem Biophys Res Commun* 146:53–60.

Ramsay RR, Singer TP (1986): Energy-dependent uptake of *N*-methyl-4-phenylpyridinium, the neurotoxic metabolite of 1-methyl-4-phenyl-1,2,3,6-tetrahydropyridine, by mitochondria. *J Biol Chem* 261:7585–7587.

Sayre LM, Singh MP, Arora PK, Wang F, McPeak J, Hoppel CL (1990): Inhibition of mitochondrial respiration by analogues of the dopaminergic neurotoxin 1-methyl-4-phenylpyridinium: structural requirements for accumulation-dependent enhanced inhibitory potency on intact mitochondria. *Arch Biochem Biophys* 280:274–283.

Schapira AH, Cooper JM, Dexter D, Clark JB, Jenner P, Marsden CD (1990): Mitochondrial complex I deficiency in Parkinson's disease. *J Neurochem* 54:823–827.

Schapira AHV, Cooper JM, Dexter D, Jenner P, Clark JB, Marsden CD (1989): Mitochondrial complex I deficiency in Parkinson's disease. *Lancet* 1:1269.

Schapira AH, Mann VM, Cooper JM, Dexter D, Daniel SE, Jenner P, Clark JB, Marsden CD (1990): Anatomic and disease specificity of NADH CoQ_1 reductase (complex I) deficiency in Parkinson's disease. *J Neurochem* 55:2142–2145.

Shoffner JM, Watts RL, Juncos JL, Torroni A, Wallace DC (1991): Mitochondrial oxidative phosphorylation defects in Parkinson's disease. *Ann Neurol* 30:332–339.

Suzuki K,Mizuno Y, Yoshida M (1990): Inhibition of mitochondrial respiration by 1,2,3,4-tetrahydroisoquinoline-like endogenous alkaloids in mouse brain. *Neurochem Res* 15:705–710.

Tanaka M, Sato W, Ohno K, Yamamoto T, Ozawa T (1989): Direct sequencing of deleted mitochondrial DNA in myopathic patients. *Biochem Biophys Res Commun* 164:156–163.

Chapter 7

The Biology of Toxic Events in Parkinson's Disease

M. B. H. Youdim

The etiology of Parkinson's disease (PD) leading to the selective degeneration of nigrostriatal dopamine neurons remains elusive. Nevertheless, in the last 30 years many biochemical hypotheses have been presented as an explanation. These have been derived mostly from some understanding of the lesioning actions of the synthetic neurotoxins 6-hydroxydopamine (6-OHDA) and MPTP (*N*-methyl-4-phenyl-1,2,3,6-tetrahydrapyridine) (Tanner and Langston, 1990) and on the nigrostriatal dopaminergic neurons and relevant systems in animal models. Although such models are not ideal, they have contributed to the understanding of some of the biological activities that can induce degeneration of nigrostriatal dopamine neurons in animals and that are also significantly altered in PD and continue to progress with the degenerative processes. It may thus be assumed that if a toxic state or a component exists, it must be sequestered within the substantia nigra (SN) (Jenner 1990, 1991; Youdim et al., 1989, 1992; Olanow, 1990).

On numerous occasions the presence of exogenously and endogenously derived neurotoxins has been proposed as the causative factor of the disease. As of now, neither an MPTP nor a 6-OHDA–like compound has been identified in the environment or within the SN (Tanner and Langston, 1990). Theoretically they could be synthesized *in vivo* and selectively accumulated within the nigrostriatal dopamine neurons. A more recent addition to this limited list of neurotoxins are the tetrahydroisoquinolines (TIQ), which can be formed endogenously from dopamine itself (Yoshida et al., 1990), probably nonenzymatically. However, the concentration of these compounds, as identified *in vivo*, are extremely

ADVANCES IN RESEARCH ON NEURODEGENERATION, I
D.B. Calne et al.

low and are far below the concentrations needed to induce degeneration of dopamine neurons and "parkinsonism" in animals (rats and monkeys) (Yoshida et al., 1990). Nevertheless, the hypothesis of a neurotoxin involvement cannot be totally ignored.

Calne and others have repeatedly stressed the multifactorial aspect of the disease. Whatever the primary cause(s) of PD, the biochemical events leading to nigrostriatal dopamine neuron degeneration must have a final common path, but not necessarily the same denominator (Calne and Takahashi, 1991). The animal studies with 6-OHDA in the early 1970s and in more recent years point to the role of oxygen free radicals in the neurotoxic action of this agent, since radical scavengers and metal chelating agents attenuated the action of 6-OHDA in peripheral catecholamine neurons (Cohen et al., 1974; Heikkila and Cohen, 1972; Sachs and Jonsson, 1975; Graham, 1978; Cohen et al., 1976). The primary biochemical lesion has been thought to be the demise of the mitochondrial electron transport system, as a consequence of oxygen free radical–induced changes in mitochondrial calcium homeostasis and membrane lipid peroxidation. Indeed, the present biochemical bases of cell death due to aging and numerous cytotoxins are closely linked to changes in cellular compartments of calcium (Boobis et al., 1989; Orrenius et al., 1989). A similar mechanism recently has been envisaged for the action of MPTP (Frei and Richter, 1989). Reports by Birkmayer et al. (1982, 1985) of longevity in parkinsonian patients on long-term monoamine oxidase (MAO)-B inhibitor (1-deprenyl) therapy as an adjuvant to *L*-dopa led these authors to propose the possible neuroprotective action of this inhibitor on the nigrostriatal dopamine neurons and the importance of oxidative stress and oxygen free radical formation due to oxidative deamination of dopamine by the action of MAO-B in the SN of PD. The consequence of dopamine deamination by MAO-B and its auto-oxidation is generation of H_2O_2. If the latter is not quenched by glutathione peroxidase, theoretically H_2O_2 in the presence of Fe^{2+} can generate substantial amounts of highly cytotoxic hydroxyl radical OH·, leading to a state of oxidative stress (Youdim et al., 1989; Olanow, 1990) (Fig. 1).

A significant increase of iron content in SN (see Youdim et al., 1989, 1993, and Jenner, 1991, for reviews), with the main elevation occurring particularly in the zona compacta (ZC), has been shown in post-mortem brain analysis and by magnetic resonance imaging (MRI) of parkinsonian patients. The well established role of transitional metals, especially iron,

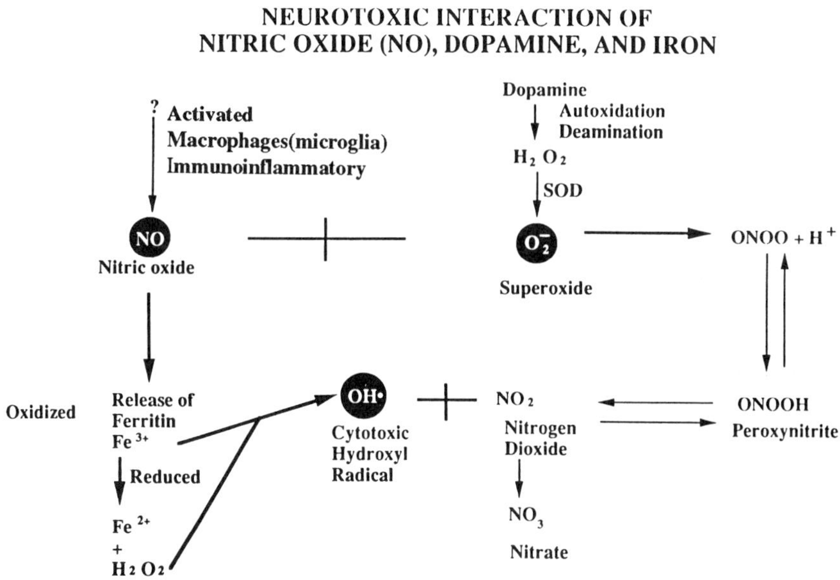

Figure 1 Metabolism of dopamine in the human brain substantia nigra via oxidative deamination by MAO-B and autooxidation, with the generation of hydrogen peroxide. In the absence of gluthathione peroxidase activity or reduced gluthathione or excess decompartmentalized iron, an interaction between hydrogen peroxide and Fe^{2+} (Fenton reaction) is envisaged. This can lead to the formation of cytotoxic hydroxyl radical (OH·) (Fenton reaction) and membrane lipid peroxidation via oxidation reduction of iron by hydrogen peroxide and melanin, respectively. The consequence of increased nigral iron (siderosis) can be tissue-oxidative stress (Youdim et al., 1989; Ben-Shachar et al., 1991b).

in the promotion of calcium-dependent oxygen radical generation resulting in oxidative stress–induced lipid membrane peroxidation and cellular death (Halliwell and Gutteridge, 1984, 1986; Halliwell, 1989) has propagated the hypothesis of iron-dependent oxidative stress in parkinsonian SN (Youdim et al., 1989). The biochemical evidence supporting it has come from diminished gluthathione peroxidase, reduced gluthathione (GSH), calcium-binding protein, and mitochondrial electron transport system complex I, together with the increase of iron, superoxide dismutase activity, ferritin, and basal membrane lipid peroxidation (see Youdim et al., 1993, for review) (Table 1). It is more than possible that the elevation of SNZC iron is secondary to some primary cause of the disease process. Nevertheless, because ionic (unbound) iron is known to be highly reactive and to be able to interact with H_2O_2 to produce highly cytotoxic hydroxyl radicals (OH·). much of the emphasis has been placed on iron

Table 1. Biochemical Changes in Parkinsonian Substantia Nigra Related to Oxidative Stress[a]

	Increased	Decreased	Unchanged
Iron	+		
Microglia	+		
Glia	+		
Astrocyte	+		
Oligodendrocytes	+		
Lewy body	+?		+
Melanized dopamine			
Neurons	+		
Ferritin	+	+	
Glutathione peroxidase		+	
Reduced gluthione		+	
Ascorbic acid		+?	
Mitochondria transport system			
Complex I		+	
Complex III		+	+
Citric cycle	?	?	?
Acotinase	?	?	?
Nitric oxide synthase		+	
Lipid peroxidation	+		
Lipofuscin	+		
Ubiquitin	+		
Melanin	+?	+	
Superoxide dismutase	+		
Copper		+	
Zn	+	+	
Mg			+
Ca^{2+}			+
Calcium binding proteins		+	
Monoamine oxidase B	+		
Monoamine oxidase A			+

[a]Taken from Schapiro et al. (1990); Mizuno et al. (1989); Riederer et al. (1989); Jenner (1991); Mizuno (1992); Sofic et al. (1988, 1991); Jellinger et al. (1990, 1992a,b). Dexter et al. (1989a,b); Youdim et al. (1993).

as one of the possible inducers of oxidative stress–dependent nigrostriatal dopamine neuron degeneration.

The evidence cited in favor of the involvement of iron recently has come from biochemical (Riederer et al., 1989) and histochemical identification and localization of increased iron in the cellular compartment of SNZC of Parkinson's disease brain (Jellinger et al., 1990). Increased iron and ferritin have been observed in glia and reactive microglia but not in the same cells of SN as in Alzheimer's disease. Iron also appears in macrophages, oligodendrocytes outside the degenerated dopamine neurons, and within the remaining melanized dopamine neurons in association with neuromelanin (Jellinger et al., 1992a,b). The persistent presence of highest brain iron in the SN confirms the ability of the brain to sequester the metal. This process could itself participate in the actual process of neurodegeneration. The results from postmortem parkinsonian SN studies, indicating an involvement of oxidative stress, have led us to investigate the neurotoxicity of decompartmentalized iron and its *in vivo* involvement in the mechanism of nigrostriatal dopamine neuron degeneration.

Biochemical Mechanism of Oxidative Stress-Induced Neurodegeneration

One of the present hypotheses concerning the etiology of Parkinson's disease suggests that the metabolically active basal ganglia can generate a substantial amount of H_2O_2 from autooxidation and deamination of dopamine by monoamine oxidase (Fig. 1). In the condition where glutathione peroxidase activity and reduced GSH are diminished in SN, as is the case in PD (Riederer et al., 1989), H_2O_2 formed could interact with ionic transitional metals (e.g., iron) (Sofic et al., 1991; Ben-Shachar et al., 1991a; Scalia et al., 1990) to generate cytotoxic oxygen free radicals of hydroxyl type (Fenton reaction), leading to oxidative stress. Such radicals theoretically could induce degeneration of nigrostriatal dopamine neurons (Halliwell, 1989). Indeed, 6-OHDA has been reported to induce its neurotoxicity via this mechanism. If this hypothesis is correct, then chelation of iron or the scavenging of oxygen radicals may be one procedure by which formation of cytotoxic oxygen radical can be prevented and the dopamine neurons protected against degeneration (Youdim et al., 1989). Iron chelation by iron chelators has been shown to reduce iron overload–induced tissue damage in the cardiovascular system and protect against ischemia–induced brain injury and a number of other oxidative stress–dependent disorders (Wolfe et al., 1985; Krause et al., 1986; Halliwell and Gutteridge, 1984, 1986; Halliwell, 1989). Youdim et al. (1989) have proposed that the SNZC of parkinsonian brain is in a state of siderosis, where the iron and ferritin accumulate in the microglia, macrophages, and astrocytes (Jellinger et al., 1990) and within the melanized dopamine neurons in association with neuromelanin (Jellinger et al., 1992a,b). At present it is difficult to validate whether the accumulated iron in ZC of parkinsonian brains is responsible for the state of oxidative stress in SN and the initiation of dopaminergic neurodegeneration (Youdim et al., 1991). Iron accumulation or release could be secondary to the neurodegenerative process. However, recent studies on animal models of PD have indicated clearly that *in vivo* decompartmentalized iron can participate in the neurodegeneration. The dopaminergic neurotoxin (6-OHDA) is thought to owe its neurotoxicity to alteration in mitochondrial Ca^{2+} homeostasis (Frei and Richter, 1986), generation of oxygen free radicals, and lipid peroxidation (Cohen et al., 1976, Heikkila and Cohen, 1972; Sachs and Jonsson, 1975; Graham, 1978; Graham et al., 1978) before or after uptake by the nigrostriatal dopamine neuron cell bodies. This was presumed to occur as a consequence of the involvement of a metal, since catecholamines can be auto-oxidized via interaction with transitional metals (e.g., iron or copper) to form semiquinones and H_2O_2, and finaly melanin. Normally, melanin, wherever it is formed in the body, is thought to function as a protective mechanism. Its neuroprotective action in the SN cannot be ignored. Indeed, dopamine melanin and tyrosine melanin inhibit formation of oxygen free radicals and lipid peroxidation (Ben-Shachar et al., 1991a; Scalia et al., 1990). However, melanin can bind (chelate) a substantial amount of ferric iron and other substances because of the numerous reactive groups it possesses (Ben-Shachar et al., 1991a). In association with Fe^{3+}, melanin promotes lipid peroxidation, and this action is further exacerbated in the presence of Ca^{2+} (Youdim et al., 1991). Indeed, it is believed that this reaction occurs as a consequence of the reduction of Fe^{3+} to Fe^{2+}, resulting in the generation of OH· and the release of iron from its binding site on melanin. It is of interest that 6-OHDA also releases iron from its binding site in ferritin; more recently its neurotoxicity has been attributed partly to this phenomenon (Monterro and Winterbourn, 1989). The ability of iron to open Ca^{2+} channels in nerve endings, alter cytoplasmic Ca^{2+} homeostasis, and induce lipid peroxidation favors this mechanism of neurotoxicity for 6-OHDA (Braughler, 1987; Youdim et al., 1991). Such a finding also is compatible with the ability of many cytotoxic agents to release ferritin iron (Kohen and Chevion, 1985). For example, lung cytotoxicity by paraquat and diquat has been attributed to a similar mechanism where these compounds completely decompartmentalize ferritin iron, without altering the structure of ferritin, and induce oxidative stress lung toxicity

(Thomas and Aust, 1986). Pretreatment in rats and mice with the iron chelator desferrioxamine (desferal) prevents the cytotoxicity of paraquat (Kohen and Chevion, 1985), whereas iron potentiates its cytotoxicity. We recently have shown that pretreatment of rats with desferrioxamine, the prototype iron chelator, attentuates the 6-OHDA–induced degeneration of nigrostriatal dopamine neurons (Ben-Shachar et al., 1991b). Thus, prevention of the degeneration of dopamine neurons by desferrioxamine pretreatment is protective against the possible release of ferritin iron by the action of 6-OHDA. Support for this concept can be cited from studies where vitamin E pretreatment also attenuates the cytotoxic action of 6-OHDA (Cadet et al., 1989). Although desferrioxamine pretreatment does not completely protect against 6-OHDA neurotoxicity, the dopamine–dependent animal behaviors and function of dopamine are normalized (Ben-Shachar et al., 1991b). Thus, compensating mechanisms known to occur when more than 80% of nigrostriatal dopamine neurons are lesioned does not operate in desferrioxamine 6-OHDA–treated animals (Zigmond et al., 1989).

Further evidence for the neurotoxic action of ionic iron has come from *in vivo* experiments in rats, where different doses of Fe^{3+} have been injected alone or in combination with desferrioxamine or lazaroids (21-amino steroid iron chelators) into the right substantia nigra (Ben-Shachar and Youdim, 1991; Sengstock et al., 1992). Three to 15 weeks after such an operation, the animal behavioral parameters examined showed that at higher iron doses ($>20\ \mu g/5\ \mu l$) there were significant ($p < 0.001$) reductions in spontaneous locomotor activity and sniffing, with increased episodes of freezing and ipsilateral turning (Ben-Shachar and Youdim, 1991). Examination of right and left striatal dopamine, noradrenaline, serotonin, and their metabolites indicated a reduction of more than 90% of dopamine content of the right striatum without an alteration in the neurotransmitters noradrenaline or serotonin or their metabolites (Ben-Shachar and Youdim, 1991; Sengstock et al., 1992). The results have been interpreted as being indicative of the selective vulnerability of nigrostriatal dopamine neurons to iron of the dopamine neurons. In intranigral desferrioxamine–pretreated rats, Fe^{3+} was not as effective in producing lesions of dopamine neurons (unpublished data). The ability of intranigral iron injection to initiate degeneration of dopamine neurons, the reduction of striatal dopamine content, and the altered behavioral responses would support the involvement of iron as one possible cytotoxin responsible for the degeneration of nigrostriatal dopamine neurons.

These parameters are persistent for at least 4 weeks after a single iron injection: Arendash et al. (1992) have reported that it can last as long as 4 months. The turnover of iron in the rat brain (Ben-Shachar et al., 1986; Dwork et al., 1990) is extremely slow and this tissue has a capacity to sequester iron to a very high degree. Thus, more than 90% of iron deposited in the brain of newborn rat (10 days old) is still present in the adult brain (80 days old) (Dwork et al., 1990). The histochemical iron analysis of Arendash et al. (1992), in which they infused iron into the rat SN, shows iron still to be present in the SN 4 months after its injection. Indeed, the iron was accumulated in the pars compacta, the nigral region in which a selective increase of iron in parkinsonian brains has been demonstrated (Sofic et al., 1991; Jellinger et al., 1990). Thus, iron can be heavily retained in the brain and may be damaging to the dopamine neurons via its selective uptake into the pars compacta cellular structures. This phenomenon may explain why a significant amount of ionic iron is present in the parkinsonian SN but it does not explain how it reaches this specific region. Even though the results with iron–induced animal parkinsonism are exciting in themselves, the selectivity of iron neurotoxicity has not been established. Other transitional metals such as Cu^{2+} and Mn could induce similar changes. Nevertheless, in PD the SN contents of these metals, unlike that of iron, are unchanged.

Dopamine and related compounds also must be considered in neuronal cytotoxicity. The studies of Cohen et al. (1975) and Graham et al. (1978) concluded that, while 6-OHDA cell killing activity depends on formation of oxygen free radicals of hydroxyl type, for dopamine and dopa, the reaction of quinone oxidation products with nucleophiles also could contribute to the neurotoxicity. In their studies they used a copper chelator to prevent the neurotoxicity; they did not consider the role of endogenous iron or an iron chelator in the reaction system. However, they suggested the formation of cytotoxins via auto-oxidation of dopamine and dopa probably to melanin. Recently, Tanaka et al. (1991) have shown that dopamine and dopa cause the death of cultured neurons in the presence of iron and not in its absence. The iron chelator desferrioxamine was able to protect the neuronal culture from death. These findings certainly support the conclusion reached by Ben-Shachar et al. (1991b) and Ben-Shachar and Youdim (1991) on the role of iron in neurodegeneration. However, if iron does contribute to the process of neurodegeneration, it must be in an ionic form, and not inert or bound to ferritin. The fact that we are unable to identify histochemically ferritin-bound iron but rather ferrous iron by Perl's stain technique (Jellinger et al., 1990) is indicative of free iron, probably bound to a small molecule. Nevertheless, it can now be generally accepted that wherever iron is stored in brain tissue, its release from the storage (bound) site can cause oxidative stress and induced cell degeneration or death (Halliwell, 1989).

In the final analysis, the validity of the experiments with iron can be measured by the notion whether similar processes are operative in Parkinson's disease as a consequence of iron accumulation in the SNZC (Sofic et al., 1991).

Brain Macrophages (Microglia), Nitric Oxide Generation, and Dopaminergic Neurotoxicity

The biochemical and enzymatic alterations outlined in Table 1 indicate that a state of abnormal toxicity, as related to oxidative stress, exists in SN, even at the time of death. Almost without exception all these measurements were carried out in whole homogenates of SN. Thus, if a toxic state or component does exist in SN, it must be present either in the whole cellular compartments (neuronal and non-neuronal) of SN or solely in the extracellular (non-neuronal) compartments and be highly sequestered. This is valid since dopamine neurons would at best represent 5% to 10% of the total cells. If the toxic state or the biochemical and enzymatic changes are solely representative of the neuronal compartment they would not be expressed, as measured in whole SN homogenate preparation. They would be dwarfed by the much higher activities expressed in the longer extraneuronal cellular compartments. Thus, there is a need for a more comprehensive analysis of the biochemical and enzymatic changes (Table 1) at specific cellular compartments using histological and immunological techniques. The abnormality of iron metabolism in PD has been thoroughly examined by us, where we have demonstrated the cellular changes of iron (Jellinger et al., 1990; Jellinger et al., 1992a,b). It is thus apparent that iron is increased in most of the cellular components of SNZC, especially the microglia (brain macrophages), and including the association of iron bound to neuromelanin in the remaining dopamine neurons of SN (Jellinger et al., 1992a,b).

Assuming that the dopaminergic neurotoxicity is the result of the action of an extraneuronally (microglia, astrocytes, or macrophages) produced cytotoxin, one must ask how are the extraneuronal cellular compartments themselves protected from it, even though major biochemical aberrations occur in PD in these cells? The answer to this question is difficult to arrive at. Nevertheless, indications are that a general cytotoxin, endogenously produced general cytotoxin and linked to abnormal iron metabolism, can be generated by the metabolic activities of macrophages, microglia, and astrocytes.

It has been known for a long time that macrophages are cytotoxic to human cells in culture when they are activiated by tumor necrosis factor or interferon-γ in combination with a second signal such as *Escherichia Coli* lipopolysacharide (Frei et al., 1987). One aspect of this cytotoxicity is a selective inhibition of certain metabolic processes in the activated macrophage effector cells (target cells) (Lancaster and Hibb, 1990). The mechanism of this metabolic cytotoxicity leading to the inhibitory states was unknown until recently. However, recent evidence suggests that although nitric oxide (NO) production has important physiological functions (Moncada et al., 1988) it can, on the other hand, be highly cytotoxic both in cellular compartments of systemic tissues as well as in the central nervous system (McCall and Vallance, 1991; Knowles et al., 1989; Bridt and Snyder, 1989). It is this latter action that has aroused most interest with regard to the neurodegenerative diseases.

Indeed, the metabolic abnormalities described in the SN of PD (Table 1) indicate that there may be a close link between the activated brain macrophages (reactive microglia), NO production, abnormal iron metabolism, and the demise of the mitochondrial electron transport system in SN of parkinsonian brains (Table 1). Indeed, Thery et al. (1991) have clearly shown that a coculturing of brain macrophages with primary neuron mesencephalic cultures results in neuron death, which can be protected against by superoxide dismutase (SOD) and catalase. They have explained these data via the action of macrophage–induced liberation of oxygen free radicals, superoxide $O\text{-}_2$, and hydroxyl radical (OH·). However, more importantly, they failed to explain what activates the macrophages to initiate the process of neuronal death. Indeed, in PD numerous reactive microglia (brain macrophages) proliferate in the SN (McGeer et al., 1988a,b). This finding has led these authors to suggest the presence of an inflammatory process, in which the brain macrophages (reactive microglia) participate in dopamine neuronal death, possibly via formation of oxygen radicals.

When activated macrophages produce nitrite and nitrate, NO and peroxynitrite are intermediate in their synthesis (Figs. 2 and 3). It is apparent that L-arginine is essential for synthesis of NO via NADPH and

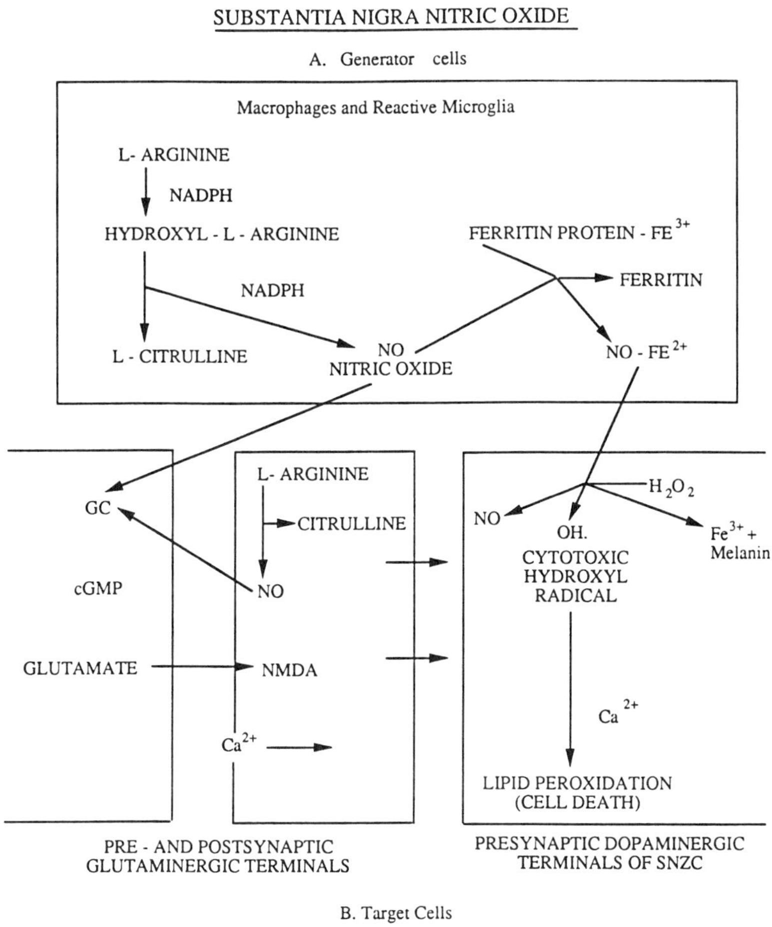

Figure 2 Formation of NO by the brain macrophages and cortical glutaminergic neurons innervating the SNZC, leading to decompartmentation of ferritin iron, formation of iron-nitrosyl moiety, and cytotoxic hydroxyl radical (OH·) in the generator cells and its effect on target cells (Garthwaite, 1991; Meldrum and Garthwaite, 1990). The consequence of these reactions can be Ca^{2+}-dependent release of glutamate at the presynaptic dopamine neurons of SNZC (Youdim et al., 1992) as an excitotoxic reaction.

calcium–dependent NO synthetase (Dawson et al., 1991a) in activated macrophages; this effect can be inhibited by NO synthetase inhibitors N-monomethyl-L-arginine and nitroarginine. One significant aspect of

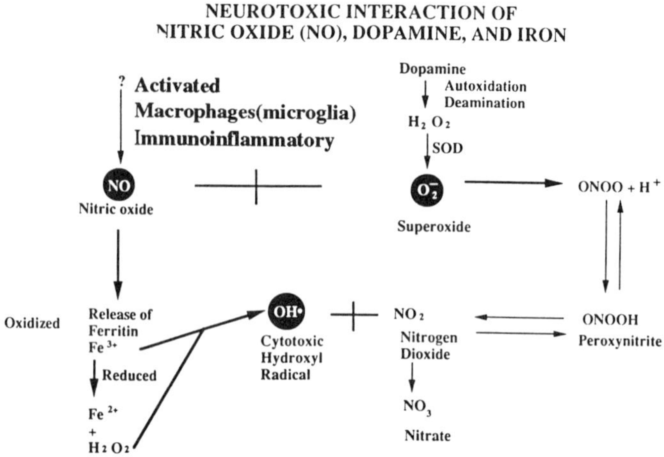

Figure 3 The possible neurotoxic interaction of nitric oxide (NO), dopamine, and iron in SNZC. This system allows the formation of iron-dependent cytotoxic hyroxyl radical (OH·) via release of ferritin iron by NO and the interaction of NO with superoxide ($O\text{-}_2$) generated by oxidation of dopamine to give highly reactive peroxynitrite. Peroxynitrite, being highly reactive, is thought to generate non–iron-dependent OH· (Beckman et al., 1992; Hogg et al., 1992; Ben-Shachar et al., 1991a).

activated macrophages, when cocultured with target cells (e.g., tumor cells or neuronal cells), is the loss of cellular iron (Hibbs et al., 1984; Wharton et al., 1988). This in turn may be associated closely with the selective inhibition of the iron-dependent mitochondrial enzymes NADH coenzyme Q reductase (complex I), succinate coenzyme Q reductase (complex II) (Granger and Lehninger, 1982), and acotinase (Drapier and Hibbs, 1986). The activities of these iron-dependent enzymes have repeatedly been reported to be decreased in the SN of parkinsonian brains (Table 1). The ability of activated macrophages to inhibit non-heme iron-dependent enzymes of the electron transport system, leading to mitochondrial inhibition (and presumably eventual cell death), may result from synthesis of NO and the direct interaction of NO on the Fe-S centers of these enzymes. This is a valid assumption since NO is a reactive oxygen radical. Indeed, iron-nitrosyl complexes in cytotoxic-activated macrophages effector cells have been observed with ESR spectroscopy (Lancaster and Hibbs, 1990). Until recently the source of this iron was unknown. Wharton et al. (1988) suggested that the iron may be liberated from iron storage protein ferritin. The studies of Reif and Simmons (1990) and our own (Youdim et al., 1993) have shown the abiity of NO (sodium nitroprusside) to release iron from ferritin *in vitro* and that the released iron catalyzes lipid peroxidation. Considering that 1 mol of ferritin can store up to 4500 atoms of iron, the release of the latter iron in free form could have a profound ability to induce oxidative stress, as is the case for the cytotoxic actions of paraquat and 6-OHDA.

Thus, there are two pathways by which activated brain macrophages can induce dopamine neuronal death, and both can be linked to NO production and free iron. First, in cerebellar and hippocampal cells glutamate can stimulate NO production, or NO produced by activated macrophages can be taken up by the presynaptic glutaminergic neurons (Garthwaite, 1991). The end result is stimulation of cyclic guanine monophosphate) (cGMP) synthesis through activation of guanylate cyclase (McCall and Vallance, 1991); in this state NO can be cytotoxic via Ca^{2+}–dependent release of excitotoxic glutamate (Garthwaite, 1991). There is a significant amount of evidence for the involvement of corticostriatal glutaminergic neurons in the pathophysiology of neurodegeneration in PD (Meldrum and Garthwaite, 1990; Turski, 1991; Olney, 1989) that also could be initiated by NO. In cultures of cortical hippocampal and striatal neurons, sodium nitroprusside (NO generator) and *L*-arginine are directly toxic to neurons and induce cell death (Dawson et al., 1991b; Vincent and Hope, 1992).

This action can be blocked with either inhibitors of NO synthetase or trapping of NO by hemoglobin by the action of the NMDA (*N*-methyl-*D*-aspartate) antagonist, MK801 (Dawson et al., 1991b). MK801 has been reported to block the neurotoxic action of sodium nitroprusside. Thus, the innervation of SNZC corticostriatal glutaminergic neurons theoretically could be involved in the dopaminergic neurotoxicity (Turski, 1991). However, convincing evidence for it is lacking.

Second, the results of studies with 6-OHDA, paraquat, diquat, and NO have demonstrated that they are all able to release iron from ferritin *in vitro* and induce lipid peroxidation and cellular death in culture, which suggests that these actions could occur *in vivo*. Indeed, iron chelators and free radical scavengers (e.g., vitamin E) *in vivo* can block the cytotoxic action of some of these ferritin-iron depleters.

With the exception of NO, the agents so far shown to release iron from ferritin are synthetic and no equivalent compounds have been identified *in vivo*. Thus, NO could be the first endogenous synthesized agent capable of decompartmentalizing iron from its protein binding site. The loss of iron from the cells via NO generation could lead to cellular injury and cytotoxicity by disrupting the cellular homeostasis of iron, which is tightly controlled and sequestered (Ben-Shachar et al., 1986; Dwork et al., 1990). Indeed, although NO by itself hardly initiates membrane lipid peroxidation as measured by malonyldialdehyde liberation, it synergistically potentiates that induced by iron or ferritin *in vitro* using rat brain cortical membrane preparations (Youdim et al., 1993). Reif and Simpson (1990) have suggested three possible ways by which these could occur: a) NO inactivates Fe-requiring enzymes important for mitochondrial respiration and DNA synthesis, b) NO may decrease the intracellular storage of iron associated with ferritin, thus preventing the cell's ability to recover the lost iron–dependent enzyme activities in the mitochondria, and c) iron released by NO could participate in the generation of oxygen free radical species resulting in propagation of irreversible damage to DNA, proteins, and membrane structures. Finally, NO itself can interact with iron or superoxide, $O\text{-}_2$, in a synergistic reaction to produce similar changes, since brain macrophages produce a significant amount of superoxide (Cotton and Gilbert, 1987) (Fig. 3). Cytotoxic–activated brain macrophages also produce superoxide (Thery et al., 1991) and superoxide is capable of releasing iron from ferritin (Samokyszyn et al., 1988; Thomas et al., 1985; Reif et al., 1989). Thus, the actions of NO synthesized by the cytotoxic–activated macrophages can be complex and are influenced by several factors, which include the rate of NO production, the availability of free iron, and the extent of scavenging superoxide by endogenous SOD, which has been reported by be decreased in parkisonian SN (Jenner, 1992). It has been suggested that SOD can increase the half-life of NO. It is more than probable that the interaction between NO and superoxide (Fig. 3) can occur *in vivo*, and macrophages can probably regulate the effect of NO by generating superoxide. The proliferation of microglia (brain macrophages) at the site of dopamine neuron degeneration in the SN as part of the inflammatory response (McGeer et al., 1988a,b) could activate the production of superoxide and NO, leading to the formation of peroxynitrite in the pathological condition as a consequence of their interaction. Indeed, peroxynitrite is a source of a strong oxidizing agent with properties identical to cytotoxic hydroxyl radical (OH·), which it also can produce (Beckman et al., 1990; Hogg et al., 1992) (Fig. 3). All these hypotheses are compatible with the pathology of Parkinson's disease as described in Table 1, where activated microphages (reactive microglia) could bring about changes in melanized dopamine neurons of SN. One important and curious feature of the macrophages-NO hypothesis of neurotoxicity is how macrophages protect themselves from the cytotoxic action of the NO they synthesize and what *in vivo* factor (e.g., inflammatory process) activates them to initiate the production of NO superoxide and peroxynitrite. Indeed, the study of macrophages, NO, and iron may hold the key to the neurodegenerative process of PD.

Conclusion

It is now apparent that the exact mechanism of nigrostriatal dopamine neuron degeneration in PD is unresolved. However, the biochemistry of SN is altered to such an extent that it indicates a state of oxidative stress, even at death (Youdim et al., 1989, 1993; Jenner, 1991). The fact that macrophages proliferate in SN of parkinsonian brains and that they can induce cellular (neuronal and non-neuronal) death via the generation of NO and release of iron from ferritin opens up new challenges for animal experimentation with animal models of Parkinson's disease. Although in this article we have concerned ourselves with the biochemical events that could lead to cell death, a genetic disposition to PD, leading to sporadic appearance of the disease in late life, cannot be excluded. A defect in the gene for enzymatic processes associated with dopamine oxidation in the mitochondria is a possibility that needs to be investigated. Thus, Hallervorden-Spatz disease, an autosomal recessive condition that, like PD,includes nigral and globus pallidus damage, is associated with iron deposition in these regions (Golbe, 1990).

References

Arendash G, Olanow CW, Dunn AJ (1992): Iron infusion model of Parkinsonism. In: *Role of Iron and Oxidant Stress in the Normal and Parkinsonian Brain.* Olanow W, Perl D, Marsden DC, Cohen G, eds. Newton: Dekker.

Beckman JS, Beckman TM, Chen J, Marshall PA, Freeman BA (1990): Apparent hydroxyl radical production by peroxynitrite: Implications of endothelial injury from nitric oxide and superoxide. *Proc Natl Acad Sci USA* 87:1620–1624.

Ben-Shachar D, Ashkenazi R, Youdim MBH (1986): Long term consequence of early iron deficiency on dopaminergic neurotransmission. *Int J Dev Neurosci* 4:81–88.

Ben-Shachar D, Eshel G, Finberg JPM, Youdim MBH (1991b): The iron chelator desferrioxamine (desferal) retards 6-hydroxydopamine-induced degeneration of nigra-striatal dopamine neurons. *J Neurochem* 56:1441–1444.

Ben-Shachar D, Riederer P, Youdim MBH (1991a): Iron-melanin interaction and lipid peroxidation: Implications for Parkinson's disease. *J Neurochem* 57:1609–1614.

Ben-Shachar D, Youdim MBH (1991): Intranigral iron injection induces behavioural and biochemical "Parkinsonism" in rats. *J Neurochem* 57:2133–2135.

Boobis AR, Fawthrop DJ, Davies D (1989): Mechanism of cell death. *TIPS* 10:275–280.

Braughler M (1987): Calcium and lipid peroxidation. In: *Oxygen Radicals and Tissue Injury*, Halliwell B, ed. Bethesda: FASEB.

Bredt DS, Snyder SH (1989): Nitric oxide mediates glutamate-linked enhancement of cGMP levels in the cerebellum. *Proc Nat Acad Sci USA* 86:9030–9033.

Cadet JL, Katz M, Jackson-Lewis V, Fahn S (1989): Vitamin E attenuates the toxic effects of intrastriatal injection of 6-hydroxydopamine (6-OHDA) in rats: behavioral and biochemical evidence. *Brain Res* 476:10–15.

Calne D, Takahashi H (1991): The origin of idiopathic Parkinsonism. In: *Parkinson's Disease: From Basic Research and Early Diagnosis to Long Term Treatment*, Rinne UK, Nagatsu T, Horowski, R, eds. Brinklaan: Medicom.

Cohen G, Heikkila RE, Allis B, et al. (1976): Destruction of sympathetic nerve terminals by 6-hydroxydopamine: protection by 1-phenyl-30-(-2-thiazolyl)-2-thiourea, diethyldithio-carbamate methianazole cysteamine, ethanol and n-butanol. *J Pharm Exp Ther* 199:336–352.

Cohen G, Heikkila RE, McNamee D (1974): The generation of hydrogen peroxide, superoxide radical and hydroxyl radical by 6-hydroxyldopamine dialuric acid and related cytotoxic agents. *J Bio Chem* 249:2447–2459.

Colton CA, Gilbert DL (1987): Production of superoxide anions by a CNS macrophage, the microglia. *FEBS Lett* 223:284–288.

Dawson TM, Bredt DS, Fotuhi M, Hwang PM, Snyder SH (1991a): Nitric oxide synthase and neuronal NADH diaphorase are identical in brain and peripheral tissue. *Proc Natl Acad Sci USA* 88:7797–7801.

Dawson VC, Dawson TM, London ED, Bredt DS, Snyder SH (1991b): Nitric oxide mediated glutamate neurotoxicity in primary cortical cultures. *Proc Natl Acad Sci USA* 88:6368–6371.

Dexter DT, Carter CJ, Wekks FR, Agid F, Agid Y, Lees AJ, Jenner P, Marsden CD (1989a): Basal lipid peroxidation in substantia nigra is increased in Parkinson's disease. *J Neurochem* 52:381–389.

Dexter DT, Wells FR, Agid Y, Lees AJ, Jenner P, Marsden CD (1989b): Increased nigral iron content and alterations in other metals ions occurring in brain in Parkinson's disease. *J Neurochem* 52:1830–1836.

Drapier JC, Hibbs JB Jr (1986): Murine cytotoxic activated macrophages inhibit acotinase in tumor cells. Inhibition involves the iron-sulfur prosthetic groups and is reversible. *J Clin Invest* 78:790–797.

Dwork AJ, Lawler G, Zybert DA, et al. (1990): An autoradiographic study of the uptake and distribution of iron by the brain of the young rat. *Brain Res* 518:31–39.

Frei B, Richter C (1989): *N*-methyl-4-phenyl pyridine (MPP^+) together with 6-hydroxydopamine or dopamine stimulate Ca^{2+} release from mitochondria. *FEBS Lett* 198:99–102.

Frei K, Siepl C, Groscurth P, Bodmer S, Swerdel C, Fontana A (1987): Antigen presentation and tumor cytotoxicity to infereron-treated microglia cells. *Eur J Immunol* 17:1271–1278.

Garthwaite J (1991): Glutamate, nitric oxide and cell-cell signalling in the nervous system. *Trends Neurosci* 14:60–67.

Giulian D, Vaca K, Noonan CA (1990): Secretion of neurotoxins by mononuclear phagocytes infected with HIV-1. *Science* 250:1593–1596.

Golbe LI (1990): The genetics of Parkinson's disease: A reconstruction. *Neurology* 40 (Suppl. 3):7–16.

Graham D (1978): Oxidative pathways for catecholamines in the genesis of neuromelanin and cytotoxic quinones. *Mol Pharmacol* 14:633–637.

Graham D, Tiffany SM, Bell WR, Gutknecht WF (1978): Auto-oxidation versus covalent binding quinones as the mechanism of toxicity of dopamine, 6-hydroxydopamine and related compounds towards C1300 neuroblastoma cells in vitro. *Mol Pharmacol* 14:644–653.

Granger DL, Lehninger AL (1982): Sites of inhibition of mitochondrial electron transport in macrophage-injured neoplastic cells. *J Cell Biol* 95:527–535.

Halliwell B (1989): Oxidants and the central nervous system: some fundamental questions. Is oxidant damage relevant to Parkinson's disease, Alzheimer's disease and traumatic injury or stroke? *Acta Neurol Scand* 126 (Suppl. No. 80):23–35.

Halliwell B, Gutteridge JM (1984): Oxygen toxicity, oxygen radicals, transitional metals and disease. *Biochem J* 219:1–14.

Halliwell B, Gutteridge JM (1986): Oxygen free radicals and iron in relation to biology and medicine: some problems and concepts. *Arch Biochem Biophys* 246:501–514.

Heikkila R, Cohen G (1972): Further studies on generation of hydrogen peroxide by 6-hydroxydopamine: potentiation by ascorbic acid. *Mol Pharmacol* 8:241–248.

Hibbs JB Jr, Taintor RR, Vavrin Z (1984): Iron depletion: possible cause of tumor cell cytotoxicity induced by activated macrophages. *Biochem Biophys Res Commun* 123:716–723.

Hogg N, Darley-Usmar VM, Wilson MT, Moncada S (1992): Production of hydroxyradicals from the simultaneous generation of superoxide and nitric oxide. *Biochem J* 281:419–424.

Jellinger K, Kinzel E, Riederer P, Youdim MBH, Ben-Shachar D (1992a): Iron, ferritin and neuromelanin in Parkinson's disease. In: *Parkinson's Disease*, in *Advances in Neurology*, Narabayashi H, Nagatsu T, Mizumo Y, Yoshida A, eds. New York: Raven Press.

Jellinger K, Kienzl E, Rumplemair G, Riederer P, Stachelberger H, Ben-Shachar D, Youdim MBH (1992b): Iron-melanin complex in substantia nigra of Parkisonian brains: An X-ray microanalysis. *J. Neurochem* 59:1168–1171.

Jellinger K, Paulus W, Grudke-Iqbal I, Riederer P, Youdim MBH (1990): Brain iron and ferritin in Parkinson's and Alzheimer's diseases. *J Neural Transm* 2:327–340.

Jenner P (1990): Parkinson's disease: clues to the cause of cell death in the substantia nigra. *Semin Neurosci* 2:117–126.

Jenner P (1991): Oxidative stress as a cause of Parkinson's disease. *Acta Neurol Scand* 84 (Suppl. 136):6–15.

Knowles RC, Palacios M, Palmer RMJ, Moncada S (1989): Formation of nitric oxide from L-arginine in the central nervous system: A transduction mechanism for stimulation of the soluble guanylate cyclase. *Proc Natl Acad Sci USA* 86:5159–5162.

Kohen R, Chevion M (1985): Protection against paraquat toxicity in mice by desferrioxamine. *Biochem Pharmacol* 34:1843–1845.

Krause GS, Kumar K, White BC, Aust SP, Weigenstein JG (1986): Ischemia, resuscitation and reperfusion mechanism of tissue injury and prospects for protection. *Am Heart J* 3:768–780.

Lancaster Jr, Hibbs JB Jr (1990): Demonstration of iron-nitrosyl complex by cytotoxic activated macrophages. *Proc Natl Acad Sci USA* 87:1223–1227.

McCall T, Vallance P (1991): Nitric oxide takes center stage with newly defined roles. *TIPS* 13:1–6.

McGeer PL, Itagaki MD, Boyes ES, McGeer EG (1988a): Reactive microglia are positive for HLA-DR in the substantia nigra of Parkinson's and Alzheimer's diseases brain. *Neurology* 38:1285–1291.

McGeer PL, Itagaki MD, McGeer EC (1988b): Expression of the histocompatibility glycoprotein HLA-DR in neurological diseases. *Acta Neuropathol* 76:550–557.

Meldrum BS, Garthwaite J (1990): Excitatory amino acid neurotoxicity and neurodegenerative diseases. *TIPS* 11:379–387.

Mizuno Y (1992): Partial deletion of substantia nigra DNA in Parkinson's disease (personal communication).

Mizuno Y, Ohta S, Tanaka M, et al. (1989): Deficiencies of complex I subunits of the respiratory chain in Parkinson's disease. *Biochem Biophys Res Commun* 163:1450–1455.

Moncada S, Radomski MW, Palmer RMJ (1988): Nitric oxide. *Biochem Pharmacol* 37:2495–2501.

Monterio HP, Winterbourne CC (1989): 6-Hydroxydopamine releases iron from ferritin and promotes ferritin-dependent lipid peroxidation. *Biochem Pharmacol* 38:4144–4182.

Olanow CW (1990): Oxidation reactions in Parkinson's disease. *Neurology* 40 (Suppl. 3):32–39.

Orrenius S, McCornkey DJ, Bellows G, Nicotera P (1989): Role of Ca^{2+} in toxic cell killing. *TIPS* 10:281–285.

Reif DW, Schubert J, Aust SD (1988): Iron release from ferritin and lipid peroxidation. *J. Biol. Chem* 280:12771–12778.

Reif DW, Simmons RD (1990): Nitric oxide mediates iron release from ferritin. *Arch Biochem Biophys* 283:537–541.

Riederer P, Reichman A, Jellinger K, Youdim MBH (1990): Reductions in Complex I and Complex III activities of mitochondria respiratory chain. *Mov Disord.*

Riederer P, Sofic E, Rausch WD, Youdim MBH et al. (1989): Transition metals ferritin, glutathione and ascorbic acid in Parkinsonian brains. *J Neurochem* 52:515–520.

Sachs C, Jonsson G (1975): Mechanism of action of 6-hydroxydopamine. *Pharmacology* 24:1–8.

Samokyszyn VM, Thomas CE, Reif DW, Saito M, Aust SD (1988): Drug. *Meta Rev* 19:283–303.

Scalia M, Geremia E, Corsaro C, Santoro C, Baratha D, Sichel G (1990): Lipid peroxidation in pigmented and unpigmented liver tissues: protection role of melanin. *Pigment Cell Res* 3:115–119.

Schapiro AH, Cooper JM, Dexter D, Clark JB, Jenner P, Marsden CD (1990): Mitochondrial Complex I deficiency in Parkinson's disease. *J Neurochem* 54:523–527.

Sengstack GJ, Olanow CW, Dunn AJ, Arendash GW (1992): Iron induces degeneration of nigrostriatal neurons. *Brain Res.*

Sofic E, Riederer P, Heinsen H, Beckman H, Reynolds GP, Hebenstreit G, Youdim MBH (1988): Increased iron III and total iron content in postmortem substantia nigra in Parkinsonian brains. *J Neurol Transm* 74:199–205.

Sofic E, Paulus W, Jellinger K, Riederer P, Youdim MBH (1991): Selective increase of iron in substantia nigra zona compacta of Parkinsonian brain. *J Neurochem* 56:978–982.

Tanaka M, Solomatsu A, Kanai H, Hirai S (1991): Dopa and dopamine cause cultured neuronal death in the presence of iron. *J Neural Sci* 101:198–203.

Tanner CM, Langston JW (1990): Do environmental toxins cause Parkinson's disease? A critical review. *Neurology* 40 (Suppl. 3):17–31.

Thery C, Chamak B, Mallat M (1991): Cytotoxic effect of brain macrophages on developing neurons. *Eur J Neurosci* 3:1155–1164.

Thomas CE, Aust SD (1986): Reductive release of iron from ferritin by cation free radicals or paraquat and other bipyridils. *J Biol Chem* 261:13064–13070.

Thomas CE, Morehous LA, Aust SD (1985): Ferritin and superoxide-dependent lipid peroxidation. *J Biol Chem* 260:3275–3280.

Turski L (1991): Excitatory amino acid antagonists and Parkinson's disease. In: *Parkinson's Disease From Basic Research and Early Diagnosis to Long-Term Treatment*, Rinne UK, Nagatsu T, Horowski R, eds. Bussum: Medicom.

Vincent SR, Hope BT (1992): Neurons that say NO. *Trends Neurosci* 15:109–114.

Wharton M, Granger DL, Durack DT (1988): Mitochondrial iron loss from Leukemia cells injured by macrophages: A possible mechanism of electron transport chain defect. *J Immunol* 141:1311–1317.

Wolfe L, Olivier N, Sallan D (1985): Prevention of cardiac disease by subcutaneous desferrioxamine in patients with thalassemia major. *N Eng J Med* 312:1600–1604.

Yoshida M, Ogawa M, Niwa T, Nagatsu T (1990): Parkinsonism produced by an endogenous substance tetrahydroisoquinoline. In: *Basic and Clinical and Therapeutic Aspects of Alzheimer's and Parkinson's Diseases*, Nagatsu T, Fisher A, Yosida M, eds. New York: Plenum Press.

Youdim MBH, Ben-Shachar D, Riederer P (1989): Is Parkinson's disease a progressive siderosis of substantia nigra resulting iron and melanin induced neurodegeneration? *Acta Neurol Scand* 126:47–54.

Youdim MBH, Ben-Shachar D, Riederer P (1993): The possible role of iron in the etiopathology of Parkinson's Disease. *Mov Disord* 8:1–12.

Youdim MBH, Ben-Shachar D, Pollard HB (1991): Iron chelators and calcium channel antagonists as inhibitors of iron induced neuronal lipid peroxidation. *Br J Pharmacol* 102:376P.

Zigmund M, Berger W, Grace AA (1989): Compensatory response to nigrostriatal bundle injury. *Mol Chem Neuropathol* 10:185–200.

Chapter 8

Markers for the Presence and Progression of Parkinson's Disease

Peter A. LeWitt

As a histopathological entity, Parkinson's Disease (PD) is easily recognized. Although its signs and symptoms are also unmistakable, PD still holds mysteries in regard to its cause and the mechanisms for its progression. Most attempts to analyze the "natural history" of PD have relied upon alterations in the clinical picture to infer what might be happening at neurochemical and synaptic levels. Since the depletion of dopaminergic neurons from the substantia nigra pars compacta can account for most features of PD, it has seemed reasonable to assume that it is the continuing loss of dopaminergic neurons which explains the advance of Parkinsonism. However, the evidence that helps us to understand the correlation between the clinical features and the underlying disease is still fragmentary. This report will discuss some recent research investigating neurochemical clues which might offer insights into the progression of PD.

One of the most intriguing issues related to the pathophysiology of the disorder is how advanced the neuronal and neurochemical change need to be before the emergence of Parkinsonian signs and symptoms. Findings in the Parkinsonian brain (whether due to the idiopathic disease (Hornykiewicz and Kish, 1986) or from MPTP given to nonhuman primates (Kopin and Markey, 1988) indicate that depletion of at least 70% of striatal dopamine precedes the onset of Parkinsonian features. There are several possibilities to explain this threshold for the occurence of Parkinsonism. One explanation might be that the striatal concentration of dopamine synthesized by nigrostriatal neurons is threefold in excess of what is actually needed for effective neurotransmission. The compen-

ADVANCES IN RESEARCH ON NEURODEGENERATION, I
D.B. Calne et al.

satory responses of the remaining dopaminergic neurons might be another factor determining the threshold. It might be that Parkinsonism becomes evident only when the neurons have reached their limit for increasing output of the neurotransmitter. In support of this concept, neurochemical analysis of the Parkinsonian brain has shown a markedly increased ratio of dopamine metabolite concentration to that of the parent compound (Hornykiewicz, 1982). Postsynaptic receptor changes in response to nigrostriatal neuronal dropout might also be determinants of the threshold for the onset of Parkinsonian symptomatology. Finally, it might be that trophic factors derived from the projections to caudate and putamen can maintain adequate neurotransmission only up to a 70% depletion of dopaminergic neurons.

With all of these possibilities, the pathological basis for progression of PD may be quite complex. The basis for the advance of the disorder be the sum of more than one mechanism. Considerations of a "natural history" for PD may be somewhat misleading in that the loss of nigrostriatal neurons may be more of a "catastrophe," rather than the orderly loss of these neurons with aging. What is probably most accurate about the concept of a "natural history" to Parkinsonism is the notion of a disorder with stereotyped features and the propensity for a slow worsening over the course of years. The clinical features and disabilities described by Parkinson, Charcot, Gowers, and other astute physicians of the 19th century are identical to the patterns of Parkinsonism encountered today. The transition between normalcy and Parkinsonism can be extremely subtle, though with unmistabable features. "So slight and nearly imperceptible are the first inroads of this malady," James Parkinson stated in his 1817 *Essay on the Shaking Palsy*, "... it rarely happens that the patient can form any recollection of its commencement." Parkinson's view nonwithstanding, it is common for patients to relate some particular stress or activity to the first occurrences of symptoms. Sometimes it is the occurrence of characteristic clinical signs that first draws attention to the onset of the disorder (such as a tremorous hand, decreased armswing, or the evolution of micrographic handwriting). The pattern by which Parkinsonism worsens is often less clearly perceived by patients, who sometimes report the evolution of this disorder in fits and starts rather than the gradual transition from mild to worse.

Epidemiological studies of PD have not provided evidence for environmental or other risk factors influencing the evolution of this disorder. In PD clinics throughout the world, the onset of symptoms is quite uniform and occurs in a similar normal distribution around the approximate mean age of 57 years (Koller et al., 1985). Because the most common scenario for PD is a slowly evolving disability in middle-aged to elderly individuals, there have been suspicions that the progressive loss of nigrostriatal dopaminergic innervation with age (McGeer et al., 1977) might be somehow related to the expression of this condition. However, there are no obvious biological events known to coincide with the usual age of PD onset. Furthermore, the diversity of ages for onset of Parkinsonism counters the notion that PD is in some way a consequence of CNS senescence. If aging conferred its own risks for the disorder, then the development of PD in the younger onset patient would likely be less severe. In fact, there is no evidence that the clinical course or brain pathology of young-onset PD takes on features different from presentations later in life (Quinn et al., 1987).

As discussed above, the presence of even the most subtle clinical features of Parkinsonism may actually be evidence for a disease process already far advanced. The duration is not known by which the first development in PD's biological progression precedes the onset of symptomatology. In the living patient, little is known of ways to confirm that Parkinsonism is actually present. PET scan studies with tracer ^{18}F-fluorodopa injections can be used to demonstrate the loss of nigrostriatal projections, but the utility of this method is quite limited (Eidelberg, 1992). A number of studies have claimed that measurements of substances found in the CSF can differentiate PD from normals (Chase, 1980). These neurochemical clues include compounds known to be present in striatal tissue, such as metabolites of dopamine and serotonin. Other substances which are altered in PD are linked only indirectly to dopaminergic neurotransmission. These include met-enkephalin, β-endorphin, and GABA, among others (LeWitt and Galloway, 1990).

One of the goals of my research has been to reexamine the validity of these biological markers for PD. This work has been conducted with specimens obtained in a standardized fashion from a group of more than 700 mildly affected Parkinsonian subjects enrolled in a multicenter clinical trial (The Parkinson Study Group, 1989a). Participants had CSF specimens collected on two occasions, thus providing the opportunity to measure changes associated with progression of Parkinsonian signs and symptoms. On each occasion that CSF was obtained, subjects were in a medication-free state. Subjects underwent repeated assessments using the Unified Parkinson Disease Rating Scale, permitting a correlative analysis of the clinical picture to measurements in CSF (The Parkinson Study Group, 1989b).

One direction has been to assess which CSF markers are specific for PD. If a particular substance differentiates PD from controls, the results from the two CSF collections can be used to determine whether this distinction serves as a *trait* or *state* marker of PD (LeWitt and Galloway, 1990). A *trait* marker, by definition, implicates the presence of PD regardless of its duration and severity. Hence, *trait* markers should be recognizable at the onset of PD, and, presumably, even before there might be any clinical expression of Parkinsonism. With these characteristics, *trait* markers might have applications in identifying PD risk factors, or those individuals who are in its earliest stages. It might be that certain *trait* markers might appear only in conjunction with particular features associated with Parkinsonism (such as the occurrence of resting tremor or cognitive impairment), and so could provide clues to link specific CNS systems to the pathophysiology of PD.

The other major class of biological markers are substances that can be correlated to one or more aspects of the PD *state*. *State* markers, thus, should appear only upon the development of Parkinsonism. In contrast to a *trait* marker (which is an all-or-none phenomenon), a *state* marker should be altered by factors regulating the expression of Parkinsonism (such as its duration or severity). It may be that certain *state* markers, although present in untreated PD, might have the potential to revert towards normal upon the use of treatments for the disorder.

The first studies in this project examined CSF concentrations of homovanillic acid (HVA) and other compounds associated with dopamine metabolism. HVA is the predominant metabolite of dopamine in man, and its concentrations in dopamine-synthesizing tissues are thought to be a representative of the neurotransmitter's rate of turnover. In a number of studies, CSF HVA concentrations differentiated PD subjects from age-matched normal controls, and a few investigations suggested that it might serve as a trait marker of Parkinsonian severity (Chase, 1980; LeWitt and Galloway, 1990). Because the dropout of dopaminergic neurons is so extensive in PD, it might be anticipated that decreased HVA concentration would be a clear indicator of even mild PD.

Our studies, which measured a wide range of CSF HVA concentrations in the mildly affected Parkinsonians (LeWitt et al., 1992a) did not find a significant difference between controls and Parkinsonians with respect to HVA concentration. Mean HVA concentration in PD CSF was 30.84 ± 17.53 ng/ml, not significantly reduced from control concentrations (32.39 ± 16.81) (LeWitt et al., 1992b). Other metabolites of dopamine, including dopamine-3-sulfate, 3-methoxytyramine, and dihydroxyphenylacetic acid (DOPAC) also could not differentiate between controls and PD subjects. Free dopamine concentrations in CSF were not decreased in PD.

We were able to detect levodopa and its derivative 3-0-methyl in CSF from these unmedicated subjects. Levodopa is a transient intermediate in the pathway of dopamine synthesis, and CSF levodopa may be derived from nigrostriatal and other dopaminergic neurons engaged in dopamine synthesis. In PD CSF, we found increases in mean (± S.D.) levodopa concentration (0.33 ± 0.50 mg/ml), to almost double that found in controls $p = 0.0002$). The mean concentration (± S.D.) of the levodopa derivative 3-0-methyldopa was also increased in the PD subjects (2.51 ± 1.10 ng/ml vs. 2.18 ± 0.72 in controls ($p = 0.04$) (LeWitt et al., 1992b). The explanation for this alteration is not known but may pertain to an increased rate of dopamine synthesis in the remaining nigrostriatal neuronal pool. Investigations are underway to determine whether this phenomenon serves as a state or trait marker.

Another observation made in the course of screening a number of CSF constituents by HPLC revealed a high correlation between CSF concentrations of HVA and xanthine (LeWitt et al., 1992b). This unexpected correlation ($r = 0.82$, $p < 0.0001$) is so great as to suggest biological relationship between the two compounds. Though neither compound differentiated Parkinsonians from controls by themselves, the xanthine/HVA ratio did so highly significantly. It may be that comparison of xanthine together with HVA concentrations eliminates factors that produce variability for measurements of CSF HVA alone. Alternatively, it may be that the relationship between the purinergic metabolite xanthine and HVA represents an as yet unrecognized neurochemical disturbance in PD. We found no change between the first and second CSF collections, and so the altered xanthine/HVA ratio is a trait marker.

Other substances present in the PD CSF specimens have been unrevealing as state or trait markers. Mean concentrations of met-enkephalinlike immunoreactivity and β-endorphinlike immunoreactivity did not differentiate Parkinsonian and controls, despite initial reports to the contrary (LeWitt et al., 1993). Biopterin, the oxidized form of the pteridine cofactor involved in dopamine synthesis, was substantially decreased in PD (mean concentration ± S.D.: 18.12 ± 3.29 ng/ml vs. 24.68 ± 6.13 in controls). There was no decline in BP concentrations with progression of Parkinsonism, nor did these values correlate to indices of clinical severity. The reason for the decreased CSF concentration of biopterin, which is highly correlated to the near-normal concentration of HVA, is unknown. However, the site of biopterin synthesis and utilization is in dopaminergic neurons, and so this trait marker for PD may provide a reflection of their decreased numbers in the Parkinsonian substantia nigra.

Several other CSF constituents are under investigation for their possible value as markers of the disease process. Compounds whose concentration changes as a function of clinical severity or duration of Parkinsonism are needed to investigate the models of Parkinsonian progression discussed above. Of particular importance is the need to differentiate the primary neurochemical features of PD from compensatory responses. The near-normal concentrations of HVA in PD may represent the output from a smaller pool of nigrostriatal neurons driven to engage in increased rates of dopamine turnover, for example. Future investigation will be directed at learning additional CSF markers for PD which might be independent of the dopamine production rate.

References

Chase TN (1980): Neurochemical alterations in Parkinson's Disease. In: *Neurobiology of Cerebrospinal Fluid*, Wood JH, ed. New York: Plenum Press, pp. 207–218.

Eidelberg D (1992): Positron emission tomography studies in Parkinsonism. *Neurol Clin* 10:421–433.

Hornykiewicz O (1982): Brain neurotransmitter changes in Parkinson's disease. In: *Movement Disorders*, Chap 4, Marsden CD, Fahn S, eds. London: Butterworth Scientific, pp. 41–58.

Hornykiewicz O, Kish SJ (1986): Biochemical pathophysiology of Parkinson's Disease. *Adv Neurol* 45:19–34.

Koller W, O'Hara R, Weiner W, Lang A, Nutt J, Agid Y, Bonnet AM, Jankovic J (1985): Relationship of aging to Parkinson's Disease. *Adv Neurol* 45:317–321.

Kopin IJ, Markey SP (1988): MPTP toxicity: implications for research in Parkinson's Disease. *Ann Rev Neurosci* 11:81–98.

LeWitt PA, Bannon M, Galloway M, Kapatos G, Matson W, McDermott M, Oakes D, The Parkinson Study Group (1993): Monoaminergic metabolism, substance P, met-enkephalin, and biopterin metabolism in Parkinson's Disease. *Neurology*, in press.

LeWitt PA, Galloway MP (1990): Neurochemical markers of Parkinson's Disease. In: *Therapeutics in Parkinson's Disease*, Koller WC, Paulson G, eds. New York: Marcel Dekker, pp. 63–93.

LeWitt P, Galloway M, Matson W, Milbury P, McDermott M, Srivastava DK, Oakes D, The Parkinson Study Group (1992a): Markers of dopamine metabolism in Parkinson's Disease. *Neurology* 42:2111–2117.

LeWitt P, Galloway M, Matson W, Milbury P, McDermott M, Oakes D, The Parkinson Study Group (1992b): CSF studies in untreated Parkinson disease: dopamine metabolism and xanthine. *Soc Neurosci Abs* 18:1248.

McGeer PL, McGeer EG, Suzuki JJ (1977): Aging and extrapyramidal function. *Arch Neurol* 34:33–35.

Quinn NP, Critchley P, Marsden CD (1987): Young onset Parkinson's Disease. *Mov Disord* 2:73–91.

The Parkinson Study Group (1989a): DATATOP: a multicenter controlled clinical trial in early Parkinson's disease. *Arch Neurol* 46:1052–1060.

The Parkinson Study Group (1989b): Effect of deprenyl on the progression of disability in early Parkinson's disease. *N Engl J Med* 321:1364–1371.

Part III

Alzheimer's Disease

Chapter 9

Definition of Alzheimer's Disease with Reference to Lewy Body Dementia

Elaine Perry and Robert Perry

Satisfactory definition of disease in relation to cure or prevention depends on the disease being a specific, unitary process with a single etiopathology. The following discussion addresses the question of whether such a deterministic definition can be achieved for Alzheimer's disease. The exmaple of Lewy body dementia is included as a focal point in the discussion.

Classical Definition

The classical or traditional definition of Alzheimer's disease is based on clinical evidence of a slowly progressive dementia including memory loss as a presenting and continuing feature and accompanied neuropathologically by the widespread distribution in the cortex of numerous senile plaques and neurofibrillary tangles—also excluding all other diseases known to be associated with dementia. Although Alzheimer's original case description was based primarily on evidence of neocortical tangles, the debate as to whether or not the disease can occur in the absence of cortical tangles has continued. Neuropathological criteria for Alzheimer's disease currently employed in Newcastle include the presence of both neocortical neuritic plaques and tangles at a mean density greater than 10 mm^2 in the four cortical lobes and the absence of other diseases such as vascular dementia (Perry et al., 1990b).

ADVANCES IN RESEARCH ON NEURODEGENERATION, I
D.B. Calne et al.

New Horizons

Recent advances in molecular genetics indicate that new definitions of Alzheimer's disease may emerge based more fundamentally on the occurrence of extensive β-amyloidosis (taking the form of dense-cored amyloid plaques). The discovery of a mutation in the β-amyloid precursor protein (βAPP) gene in familial Alzheimer's disease (Goate et al., 1991) has been interpreted by some as evidence that β-amyloidosis is the core pathogenic event in all cases of the disease. The hope that molecular genetics can provide the ultimate definition of Alzheimer's in terms of βAPP abnormal processing is however not yet realized. Although several different mutations have been reported in the βAPP locus on chromosome 21 and also on other chromosomes (14, 19 and 22, to date). Moreover transgenic models based on abnormal βAPP-expression have not yet mimicked the pathology of Alzheimer's disease in terms of neuritic plaques and neurofibrillary tangles.

The hypothesis that β-amyloidosis is the central pathological event in Alzheimer's disease also needs to be evaluated critically in the face of a number of other problems. Thus, amyloidosis occurs not only in a broad range of other diseases (including Parkinson's—see below) but also, more importantly in almost all nondemented elderly people. Moreover, a range of phenotypic expressions of APP gene abnormalities include such non-Alzheimer disease processes as hemorrhagic amyloidosis (for discussion, see Kay, 1991). Presumably numerous variations in genetically determined and post-transcriptional (experimental or environmental) events involved in abnormal APP processing are possible that, in combination with other genetic variations, are manifest as a range of differing pathological and clinical features. Ultimately, the major contribution of molecular genetics to the understanding of Alzheimer's disease is likely to be the unquivocal evidence of disease heterogeneity.

Subgroups

One of the challenges in attempting to define Alzheimer's disease is the need to accommodate the existence of distinct subgroups or, alternatively, devise independent classifications of these atypical forms. Apart from genetics, numerous atypical forms of Alzheimer's disease, based on neuropathology, have been reported, including those with Lewy bodies, few or absent tangles (and even no plaques), and mixed pathologies (e.g., infarction or tunors, especially in old age). Further subgroups have been identified on the basis of regional variations in cortical versus hippocampal pathology, the occurrence of congophilic angiopathy, or extent of neuron (e.g., locus ceruleus) loss. Clinical subgroups include familial, presenile, cases with marked extrapyramidal features, cases (see above) with presenting visuospatial or psychotic features, and, in terms of treatment, drug responders versus nonresponders. Molecular genetics may assist in defining pathogenic factors leading to the development of these different forms. If it transpires that there are numerous variations in the localization and the nature of the gene defects or in nongenetically determined abnormalities in phenotypic expression, then the term Alzheimer's disease will become more appropriately a collective term for a spectrum of disorders rather than a unitary process.

Lewy Body Dementia

The problems of definition in Alzheimer's disease are exemplified in the instance of dementia of Lewy body type. Examination of this disease from clinical, pathological, and neurochemical viewpoints suggests that a more precise definition of Alzheimer's disease would place Lewy body dementia in a separate category. Dementia associated with cortical Lewy bodies has been variously classified as diffuse Lewy body disease, Lewy body dementia, senile parkinsonism, senile dementia of Lewy body type, and the Lewy body variant of Alzheimer's disease (for recent references, see Perry et al., 1990b; Hansen et al., 1990; Byrne et al., 1991).

Clinical Features

The presenting or principal clinical features of this type of senile dementia of Lewy Body type (Table 1) usually are distinct from that of Alzheimer's disease, including fluctuating confusion, psychosis (e.g., hallucinations or depression), generally mild extrapyramidal features which may be severe and life threatening following neuroleptic medication, less severe memory loss and dementia, and a shorter duration of illness. Although different from classical cases of Alzheimer's, the syndrome bears little resemblance to the motor disorder that characterizes Parkinson's disease with its primarily extrapyramidal as opposed to cognitive deficits.

Table 1. Clinical Features of Alzheimer and Lewy Body–Type Dementia

	Alzheimer's Disease	Lewy Body Dementia
Case numbers	37	21
Age at disease onset (ys)	75 ± 4	73 ± 6
Mental test score		
(Blessed et al., 1968) – onset	15 ± 9	25 ± 7
– terminal	5 ± 8	13 ± 9
Hachinski score	2.3 ± 2.0	5.6 ± 2.2
Extrapyramidal features (mild)*	0.9 ± 0.9	2.4 ± 1.5
Fluctuating confusion	4 (11%)	19 (90%)
Visual hallucinations	5 (14%)	11 (52%)

* Following neuroleptic medication a proportion of patients with Lewy Body, but not Alzheimer dementia develop irreversible Parkinsonism leading to reduced survival (McKeith et al., 1992).

Neurochemistry

Neurochemically, Lewy body dementia shares some of the features of both Parkinson's and Alzheimer's disease (Perry et al., 1990c,d). A loss of striatal dopamine (and HVA), extensive neucortical cholinergic enzyme reductions, and muscarinic receptor supersensitivity more closely resemble Parkinson's; an extensive archicortical (e.g., entorhinal gyrus) cholinergic abnormality more closely resembles Alzheimer's. Although such variations in transmitter pathology are likely to relate differences in symptoms, a useful definition of Alzheimer's disease or Lewy body dementia, however, is not immediately apparent based on neurochemical these transmitter abnormalities.

Neuropathology

Neuropathologically, the presence of Lewy bodies in both the brain stem and cortex distinguish the group from most cases of "classical" Alzheimer's disease and is similar—although differentially distributed—to that seen in Parkinson's disease (Table 2). Whereas tangles generally are absent or sparse, plaque densities in most are as high as those seen in Alzheimer's disease (although there is a minority of around 20% with no more amyloidosis than is apparent in age-matched normal individuals and the Alzheimer-type pathology in many cases resembles that found in many elderly Parkinson cases). In the brain stem, substantia nigra

Table 2. Neuropathological Features of Alzheimer's Disease, Senile Dementia of Lewy Body Type, and Parkinson's Disease

	Control	Alzheimer's Disease	Lewy Body Type Dementia	Parkinson's Disease
Case numbers	14	14	14	14
Age	78±6	81±5	81±6	76±5
Neocortical plaques (mean/mm^2)	3.0±3.8	16.2±6.6	16.4±10.4	7.2±7.1
Neocortical tangles (mean/mm^2)	0.01±0.02	29.9±19.6	1.5±4.3	0.06±0.14
Archicortical Lewy bodies (mean/cm^2)	ND	ND	50.8±35.7	17.0±17.0
Substantia nigra (neurons/20 μm)	561±141	540±108	328±152	173±108
Locus ceruleus (neurons/20 μm)	101±27	32±22	29±22	17±12

ND, not determined.

neuron loss (40% reduction of age-matched control numbers) is not as great as that in Parkinson's (over 70%), presumably reflecting the mild as opposed to severe extrapyramidal disorder. The neuropathological profile in general is thus related more closely to Parkinson's than Alzheimer's disease—the latter being distinguished by the absence of Lewy bodies but high density of tangles in neo- and archicortex. A more useful definition of "classical" Alzheimer's disease may thus be dementia accompanied by β-amyloidosis and neocortical tangles, in equivalent densities.

On balance, the neuropathology (including transmitter defects) suggests that Lewy body dementia may belong to a non-Alzheimer spectrum of disease that includes Parkinson's disease—a conclusion that, if verified, has important implications for the epidemiology of Parkinson's disease since current prevalence data suggest that the incidence of Lewy body dementia in old age may exceed that of Parkinson's disease.

Genetic

Evidence against the above proposal includes the presence of Lewy bodies in one case affected by the missense mutation at codon 717 of the APP (Goate et al., 1991). Moreover, in a minority of middle-aged patients with Down's syndrome, Lewy bodies also are apparent (Raghavan et al., in preparation). However, Lewy bodies are absent in the great majority of patients with Alzheimer's disease defined by concomitant β-amyloidosis and neurofibrillary tangles (Table 2). Moreover, a distinct etiology for Lewy bodies and amyloid-cored plaques is suggested by recent immunohistochemical findings (Wisniewski et al., 1991). Lewy bodies react with an antibody against amyloid in familial amyloidosis, Finnish type, which has significant sequence homology with gelsolin (Haltia et al., 1990), indicating that Lewy bodies contain gelsolin or a gelsolin variant epitope (gelsolin codes to chromosome 9, not 21).

Natural Aging

Definition of diseases such as Alzheimer's and Lewy body dementia that occur predominantly in old age may be clarified further by consideration of natural age-associated neurodegenerative "pathology" in the normal population. Diseases of old age associated with dementia may be divided into two categories: those characterized by specific features not evident as part of the natural aging process and those indistinguishable in any respect, except degree, from natural aging (Perry and Perry, 1988). Whereas Alzheimer's disease falls into the latter category, Lewy body dementia (together with Parkinson's disease and a host of rare diseases such as Pick's, Huntington's, Creutzfeldt Jakob, etc.) falls into the former.

The prevalence of Lewy bodies in the normal elderly individual is considered to be low (estimates range from 3% to 16%, the upper values being perhaps artifactually high on account of the inclusion of cases of psychiatric Lewy body disease; for discussion see Perry et al., 1990a). In the elderly population as a whole, Lewy bodies may occur in under 10%, being accounted for by cases of Parkinson's (1–2%), Lewy body dementia (1–2%), and neurologically or psychiatrically normal individuals (2–5%). It seems reasonable to conclude from this estimate that the presence of Lewy bodies in otherwise normal individuals signifies preclinical cases of Lewy body disease as opposed to a nonspecific age-related change, and that Lewy body–associated disorders are part of specific disease processes.

In contrast to Lewy bodies, neocortical senile plaques (but not tangles) are evident in the vast majority of nondemented old people with prevalence figures of greater than 90% reported in the 9th and 10th decades. As has been argued previously (Perry and Perry, 1988; Lishman, 1991), amyloidosis must occur independently of Alzheimer's disease or nearly the entire human race would develop the disease if they survived long enough. One resolution of this dilemma is the hypothesis that a process of "alzheimerization" occurs to a greater or lesser extent in all individuals with increasing age, and whether or not clinical symptoms of disease are manifest depends on the extent of Alzheimer-type pathology, which in turn reflects a broad range of interactive genetic and environmental factors. A more satisfactory resolution of the dilemma may be the inclusion of substantial neurofibrillary tangle formation (equivalent in density to that of neuritic plaques) in the neocortex (not apparent in natural aging) as an essential feature of typical Alzheimer's disease.

A "Majority" Definition

The question arises as to whether a unitary definition of Alzheimer's disease can be found that will satisfy all those involved in the disease—the clinician requiring a diagnosis, prognosis, a disease-specific marker, and useful treatment; the neuropathologist seeking classification based on regional distribution of specific neurodegenerative features; the neurochemist and epidemiologist requiring uniform groups in which to establish disease effects and associations; and the molecular geneticist pinpointing a gene locus and protein abnormality that explain the primary pathogenesis of the disease. An optimistic viewpoint is that these requirements may be broadly satisfied in most (especially sporatic) cases of Alzheimer's disease defined by β-amyloidosis, associated with neuritic plaques, and neurofibrillary tangle formation and probably also neuropil threads. However, variations in genotype, the complexity of phenotypic expression, or perhaps interaction with other diseases gives rise to a minority of cases in which this single definition is not satisfactory. As and when an *in vivo* disease "marker" becomes available and routine population screening possible, this definition may then identify most at-risk individuals for preventive therapy.

Overview

This "majority" definition of Alzheimer's disease (which does not radically differ from the original "classical" definition) still fails to pinpoint any qualitative difference between the disease and aging. If the ultimate objective of definition is an understanding of the disease etiology with a view to its eradication, a less optimistic viewpoint is that prevention of amyloidosis will have to strike at the heart of cerebral aging itself. New markers differentiating natural aging from Alzheimer's disease may emerge but the pragmatist still has, as a major objective, the search for therapeutic strategies to alleviate disease symptoms as they arise and to prevent further deterioration.

Summary

In summary, a current definition of Alzheimer's disease based simply on the occurrence of β-amyloidosis as the primary etiopathological event, manifest as amyloid-cored plaques, leaves critical issues unresolved. A more useful definition, applicable for most cases, includes the presence of substantial neocortical tangle formation and neuropil disruption or 'threads.' This would distinguish Alzheimer's disease from natural aging and from such diseases as Lewy body dementia (discussed above).

The following questions arise for discussion:

1. Should the term Alzheimer's disease be replaced with "β-amyloidosis" or "β-amyloid dementia"?
2. Should the definition include neocortical tangles (and neuropil threads) in addition to dense-cored amyloid plaques?
3. Is Alzheimer's a unitary disease or a spectrum of disorders?
4. Do subgroups represent atypical genotype, phenotype, coexisting diseases or are they separate diseases?
5. Will new markers distinguish Alzheimer's disease from aging and other β-amyloid–associated disorders?
6. If "alzheimerization" is a ubiquitous process, should definition still depend on traditional clinical and pathological indices?
7. Should definition depend on regional anatomy (an issue not addressed in detail in the above discussion)?

As the subject of Alzheimer's disease increases almost exponentially in complexity, it can only be hoped that, in accordance with modern chaos theory, some order will shortly emerge.

References

Byrne EJ, Lennox GG, Godwin-Austen RB, Jefferson D, Lowe J, Mayer RJ, Landon M, Doherty FJ (1991): Dementia associated with cortical Lewy bodies: proposed clinical diagnostic criteria. *Dementia* 2:283–284.

Goate A, Chartier-Harlin MC, Mullan M, et al. (1991): Segregation of a missense mutation in the amyloid precursor protein gene with familial Alzheimer's disease. *Nature* 349:704–706.

Haltia M, Prelli F, Ghiso J, Kiuru S, Somer H, Palo J, Frangione B (1990): Amyloid protein in familial amyloidosis (Finnish type) is homologous to gelsolin, an actin binding protein. *Biochem Biophys Res Commun* 167:927–932.

Kay DWK (1991): Some developments in the genetics of Alzheimer's disease and related disorders. *Dementia* 2:195–200.

Lishman WA (1991): The evolution of research into dementia. *Dementia* 2:177–185.

McKeith IG, Perry RH, Fairbairn AF, Perry EK (1992): Neuroleptic sensitivity in patients with senile dementia of Lewy body type. *Brit Med J* 305:673–678.

Perry EK, Perry RH (1988): Aging and dementia: some neurochemical and neuropathological comparisons. *Life Sci Res Rep* 43:213–230.

Perry RH, Irving D, Tomlinson BE (1990a): Lewy body prevalence in the aging brain: relationship to neuropsychiatric disorders, Alzheimer type pathology and catecholaminergic nuclei. *J Neurol Sci* 100:223–233.

Perry RH, Irving D, Blessed G, Fairbairn A, Perry EK (1990b): Senile dementia of Lewy body type: a clinically and neuropathologically distinct form of Lewy body dementia in the elderly. *J Neurol Sci* 95:119–139.

Perry EK, Marshall E, Perry RH, Irving D, Smith CJ, Blessed G, Fairbairn AF (1990c): Cholinergic and dopaminergic activities in senile dementia of Lewy body type. *Alz Dis Assoc Disord* 4:87–95.

Perry EK, Marshall E, Kerwin J, Smith CJ, Jabeen S, Cheng AV, Perry RH (1990d): Evidence of a monoaminergic:cholinergic inbalance and relation to visual hallucinations in Lewy body dementia. *J Neurochem* 55:1454–1456.

Wisniewski T, Haltia M, Ghiso J, Frangione B (1991): Lewy bodies are immunoreactive with antibodies raised to gelsolin related amyloid – Finnish type. *Am J Pathol* 138:1077–1083.

Chapter 10

Morphology of the Aging Brain and Relation to Alzheimer's Disease

Kurt A. Jellinger

The influence of biological aging on the brain includes a variety of changes that show considerable variability in their degree between the different regions. The age-related morphologic changes of the brain include a) atrophy with decline in volume, weight, numbers, and size of neurons, b) changes of neuronal cytoskeleton, deposition of lipofuscin and of β-amyloid, and c) changes of astroglia and of the blood–brain barrier. Here, some major morphological changes of the aging human brain are updated (for review see Kemper, 1984; Rogers and Staren, 1987; Coleman and Flood, 1987; Ulrich, 1988; Jellinger, 1989; Haug, 1991) and are compared with those in Alzheimer's disease (AD).

Brain Weight and Volume

The human brain undergoes reduction in volume and weight with age that is, however, rather mild when corrected with regard to secular acceleration of the body length (Haug, 1985). Brain weight remains fairly stable until the age of 60 to 65 years, whereas in individuals over age 80 years, in comparison with younger individuals, a decrease ranging from 3% to 18% is seen. The decline in brain weight with age is significant but the onset of reduction of the brain mass is unclear. It may be around the age of 65 years, with mild increase after 80 years. The volume of the brain in the 8th decade is reduced by 6% to 10% versus the third decade, but shows considerable variations between the different brain regions (Table 1). In the 9th to 10th decade, the total brain volume is reduced by 6% to

ADVANCES IN RESEARCH ON NEURODEGENERATION, I
D. B. Calne et al., Editors

10% with a decrease of the total cortex by 2% and of the white matter by 11% (Haug, 1991).

Table 1. Reduction of Brain Weight and Volume with Age

Region	Age Range	Percent Change	Reference
Gross brain weight	19–94	-3-11	Rogers & Styren, 1987;
	24–100	-18	Terry et al., 1987a;
Brain volume	20/40–80	-6	Haug, 1991
Volume total cortex		-2	
Volume white matter		-11	
Volume frontal cortex		-10	
Volume parietal cortex		0	
Volume hippocampus	21–91	0	See Rogers & Styren, 1987;
Volume corpus collosum	20/40–80	-16.6	Haug, 1991;
	30–70	-12	Weis et al., 1991; and
Volume striatum	20/40–80	-12	Haug, 1991
Volume globus pallidus		-19	
Volume thalamus		-16	
Cortical thickness			
Midfrontal	20/40–70	-15	Terry et al., 1987a
Superior temporal		-9	
Inferior parietal		-13	

The age-related changes are more prominent in the frontal lobe, which shows a 10% reduction in volume and 15% reduction in cortical thickness than in the temporal and parietal lobes. The corpus callosum is decreased in volume by 12% to 17%, with accentuation in the anterior two-fifths because of the reduction of fronto-temporal interhemispheral fibers (Weis et al., 1991). Although the hippocampus and parietal cortex show no definite age-related reduction in volume, the basal ganglia shows a variable decrease ranging from 12% to 19% (Table 1). Decrease in brain volume is correlated with an increase of the outer and inner cerebrospinal fluid (CSF) spaces visualized by modern neuroimaging methods (CT and MRI) that, however, may not be correlated with the degree of cognitive and intellectual impairment with age, and severe brain atrophy is observed in 7% to 10% of nondemented aged subjects

(Jellinger, 1989). In AD reductions in brain weight and volume are much more severe and show different patterns than those in normative aging (Table 2).

Table 2. Neuronal Loss in Cortical Areas with Aging (70–90 v. 15–30 yrs)

Region	Percent Reduction	Reference[a]
Superior frontal gyrus	49	Brody, 1970
Precentral gyrus (area 4)	25	Brody, 1955
	49 L	Henderson et al., 1980
Mid frontal gyrus	41 L	Terry et al., 1987a
Cingulate gyrus (area 24)	15	Anderson et al., 1987a
Temporal lobe	45	Colon, 1972
Superior temporal gyrus (area 22)	57	Brody, 1955
	52 L	Henderson et al., 1980
	47 L	Terry et al., 1987a
Mid temporal gyrus (area 20)	18	Brody, 1955
	36 L	Henderson et al., 1980
Postcentral gyrus (area 3)	13	Brody, 1955
	53 L	Henderson et al., 1980
Inferior parietal gyrus	45 L	Terry et al., 1987a
Parietal cortex (area 7)	0	Haug et al., 1983
Occipital cortex, area 6	>10	
area 21	>10	
area 17	0	
area 18	31	Brody, 1955
Visual cortex	48	Devarney et al., 1980
Cerebellum, Purkinje cells	25	Corsellis, 1977

[a] For refs. see Jellinger (1989); L = large neurons.

Neuronal Counts in Cerebral Cortex

Increasing interest in both the normal neurobiology of aging and of AD has raised the question of changes in cortical and subcortical neuronal populations with aging. Numerous studies have demonstrated that aging does not affect all parts of the brain equally but shows considerable local variability. In contrast to previous studies suggesting considerable rates of neuron losses (37% to more than 50%) in human cerebral cortex (Brody,

1955; Colon, 1972), more recent (semi)automated computerized morphometric studies documented that total neuronal population in frontal, temporal, and parietal cortex is not significantly changed with age. However, there is a severe loss of large neurons (over 90 μm^2) ranging from 35% to 47% with more severe involvement of the frontal and temporal lobes, whereas the parietal cortex, severely affected in AD, is less involved in normal aging (Terry et al., 1987a). Total numbers of neurons, neuronal density, and the percentage of cells per area in most cortical regions are all unchanged, because of an increase with age in the numbers of small neurons and glia, particularly in midfrontal and superior temporal areas (Terry et al., 1987a). The salient age-related change is shrinkage of large neurons between 10% and 35% , with consequently increasing numbers of small neurons. Aging predominantly involves the frontal lobes (motor association area 65 with reduction in neuronal perikaryal size by 30% to 35%), the psychosocial area 11, and the superior temporal area 20 with volume reduction of neurons by 10% to 30%, whereas the visual area 17 shows no or only mild decrease in neuronal size but increased neuron density by 10% late in life (Haug, 1991). On the other hand, the astroglia shows an age-related increase ranging from 32% in parietal cortex to 76% in midfrontal cortex, causing negative correlation with age of neuron-glia ratios (Terry et al., 1987). The comparatively stable numbers of neocortical neurons in normal aging are in contrast to the extensive neuronal depletion in AD ranging from 40% to 60% associated with up to a 400% increase in astroglia (Table 3).

The hippocampus is one of the brain regions most heavily involved by AD pathology resulting in impairment of memory or cognition, but data on its affection in normal aging are controversial (Table 4). While older manual cell counts revealed neuronal losses of 20% to 30% in total hippocampus, subiculum, and H 1–3 sectors with one up to 6% cell loss per decade, more recent automatized studies showed only mild or no significant effects of age on pyramidal neuron density in any of the fields of hippocampal formation (Brown and Cassell, 1980; Davies et al., 1992). Although the pyramidal band width decreases with age in the prosubiculum, there is no effect of age in any other region of the hippocampus. The extent of dendrites in CA 1 and subiculum also was found to be stable in normal aging (Hanks and Flood, 1991; Flood, 1991). These data suggest that reduction of both pyramidal cell density and dendritic extent do not occur with age in the absence of AD pathology. This is in contrast to a significant cell decrease in AD ranging from 19%

Table 3. Quantitative Morphology of Alzheimer's Disease

Type of Change	Reduction in Percent
Brain weight	7.5–18
Hemispheral volume	13–18
Ventricular volume	35–55
Weight cerebral cortex	
Inferior parietal	45–58
Temporal	20–45
Amygdala	47
Thickness of total cortex layers	10–35
Depth of temporal cortex[a]	50
Cross-sectional area (mm^2)[b]	
Temporal lobe	35–50
Frontal lobe	30
Insular cortex	25–28
Parietal lobe	21–28
Corpus callosum	30–39
Hippocampus	28
Amygdala	38
Caudate, thalamus	26
Putamen	14–22
Numbers of neurons (90μm)	
Midfrontal cortex	25–33
Precentral gyrus[a]	70
Superior temporal gyrus	22–40
Hippocampus	35–57
Occipital cortex[a]	45–50
Inferior parietal	41
Dendritic spines	
Frontal cortex	17
Temporal cortex	36
Synapses[c]	
Midfrontal cortex	34–47
Inferior parietal cortex	51–59
Superior temporal cortex	35–40
Astroglia, frontal cortex	+400
Subcortical nuclei, neuron loss	
Caudate nucleus[a]	12
Putamen[a]	3
Pallidum[a]	9
Subthalamic nucleus[a,d]	10–26
Anterior thalamus[a]	22
Mammillary body	20
Nucleus basalis of Meynert[a] (lit.)	58 15/33–90
Locus ceruleus	30–85
Dorsal raphe nucleus	36–77
Substantia nigra, zona compacta	6–48
Pedunculopontine nucleus	29–34
Supraoptic nucleus	65
Dorsal vagal nucleus	44
Thalamic nuclei[d]	37–58

[a] for refs. see Jellinger (1990); Miyoshi and Sato (1991).

[b] Mann et al. (1991).

[c] Masliah et al. (1991).

[d] Xuereb et al. (1991).

in presubiculum to 44% in the CA 1 region and subiculum (Table 4) associated with severe regression of dendritic extent (Hanks and Flood, 1991; Flood, 1991).

Table 4. Neuronal Loss in Hippocampus in Aging and Alzheimer's Disease

Region	Age Range	Percent	Cell loss	Reference
Total hippocampus	47/89	27	(Over age 50)	Ball et al., 1977[a]
Subiculum	19/77	29		Shefer (1977)[a]
H1–H3	21/91	19–25	Pyramidal cells	M. Dam (1979)[a]
		15	Granule cells	
Subiculum	69/95	14	(Over age 70)	Anderson et al., 1984[a]
Sector H 1	15/96	3.5	Per decade	Miller et al., 1984[a]
		29	(Over age 15)	
	10/97	6.2	Per decade after age 50	Mann et al., 1985[a]
CA1–4	49/77	0	Total pyramidal cells	Brown & Cassell, 1980[a]
	4/98	3.8	Per decade	Mann et al., 1986[a]
CA4		25	(Over age 65)	
CA1–4	6/87	0	(All sections)	Davies et al., 1992
For comparison: Neuronal loss in Alzheimer's disease				
Total hippocampus	47			Ball, 1977[a]
Subiculum	39			Doebler et al., 1987[a]
Sector H1	22			Ince et al., 1991
Sector H1	28			
subiculum	44			Davies et al., 1992
CA1	41			
Prosubiculum	39			
Presubiculum	19			
CA3	17			

[a] For refs. see Jellinger (1989).

Pyramidal loss in these regions affects both the intrinsic hippocampal circuitry and connections to cortical and subcortical areas. Cell loss in the entorhinal cortex, the origin of the glutamatergic pathway connecting the hippocampus with most of the isocortex, will cause isolation of this region from the rest of the brain and thus is considered a major cause of memory defects characteristic of AD (Hyman et al., 1986; Braak and Braak, 1990; Brady and Mufson, 1991). Similar results have been reported for the *amygdaloid* complex that is integrated in the Papez memory loop (Unger et al., 1988; Scott et al., 1992). Aged controls show only mild reduction of nicotinamide adenine dinucleotide phosphate diaphorase (NADPH-d)–positive large cells in the amygdala, whereas a significant decrease is seen in the superior temporal gyrus. In AD, a significant atrophy of large NADPH-d–positive neurons cross-reacting with somatostatin and increased numbers of small-sized neurons are seen in amygdala and superior temporal cortex, whereas no changes are observed for neuropeptide Y-immunoreactive neurons in both aging and AD (Unger and Lange, 1992). In the human olfactory bulb, there is linear decrease of the mitral cells with age (from 51,000 at the age of 25 years to 14,500 at 95 years) (Bhatnagar et al., 1987; Struble and Clark, 1992) with occurrence of neurofibrillary tangles in 47% of individuals over 60 years (Kishikawa et al., 1990), whereas dystrophic neurites in the olfactory epithelial cells suggested to be specific for AD (Talamo et al., 1989) are frequently observed in middle-aged individuals (Trojanowsky et al., 1991).

Neuronal Counts in Subcortical Nuclei

In the brain stem, neuronal populations are unaffected by age in a number of regions up to the 9th decade, e.g., inferior olivary nucleus, cerebellar dentate nuclei, cranial nerve nuclei IV, VI and VII, ventral cochlear nucleus, etc.). For other subcortical nuclei (Table 5), variable neuronal depletion has been reported that correlates with biochemical findings in some specific neuromediator systems (for reviews see Coleman and Flood, 1987; Jellinger, 1989, 1991).

Table 5. Neuronal Loss in Subcortical Nuclei

Region	Age	Percent Loss	Reference
Putamen	16/90	30	Bugiani et al., 1979
		0	Pesce and Reale, 1987
Substantia nigra	8/92	34	McGeer et al., 1966
Zona compacta		47	Mann et al., 1984
Dorsal tier		45	Fearnley and Lees, 1991
Ventral tier		15	(5% per decade)
Locus ceruleus	15/104	47	Vijayashankar & Brody, 1979
		24	Mann et al., 1984
		70	German et al., 1988; 1992
Dorsal raphe n.	15/92	7	Mann et al., 1984
Dorsal motor X n.		30	
Inferior olive	30/80	20	Meier-Ruge et al., 1978

[a] For refs. see Jellinger, (1990).

The *striatum* shows a decrease in total volume of 12% but no significant neuronal losses with age, although the cell density increases (Haug, 1991). The striatonigral dopaminergic system shows mild damage with neuronal losses in the substantia nigra zona compacta (SNZC) particularly in the dorsal tier (average 45% cell loss) that is better preserved in Parkinson's disease (PD), while the ventral mesostriatal tier of SNZC projecting to the ventral parts of the ventral parts of the striatum and preferentially damaged in PD shows only mild neuronal loss (average 15%) in normal aging (Fearnley and Lees, 1991). These changes correlate with only mild age-related decrease of striatal dopamine and dopamine receptors (Hornykiewicz et al., 1989; DeKeyer et al., 1990). By comparison, in AD neuronal loss in SNZC ranges from 8% to 48% (see Jellinger, 1989, 1990), with more severe depletion of pigmented neurons in the medial part of the lower SNZC (38–51%) than in the medial part of the upper SNZC, with neuronal loss ranging from 21% to 43% (Uchihara et al., 1992).

The noradrenergic *locus ceruleus* shows a linear decrease of neuron numbers after the age of 65 years ranging in total from 24% to 54%, with predominant damage to rostral parts projecting to neocortex and hippocampus, and no or very few lesions in caudal parts of the nucleus projecting to the cerebellum and spinal cord (Zweig et al., 1988, Chan-Palay and Asan, 1989). More severe cell losses are seen in AD ranging from 40% to 80% with similar topographic patterns, suggesting retrograde degeneration due to primary damage to cortical target areas (Chan-Palay and Asan, 1989; Jellinger, 1991a; German et al., 1992). The most severe changes are seen in demented AD subjects with depression (Zweig et al., 1988) who have a 10 to 20-fold decrease in cortical norepinephrine levels (Zubenko et al., 1990).

The serotonergic *raphe nuclei* show little variation with age when studying either the number of PH-8 (an antibody cross-reacting with tryptophan hydroxylase, the synthesizing enzyme of serotonin) neurons (Halliday et al., 1992) or all cells in the dorsal raphe nucleus showing the average cell loss of 7% (Table 6). Conversely, in AD, the depletion of large neurons ranges from 30% to 70% (Jellinger, 1991; Halliday et al., 1992; Förstl et al., 1992; Aletrino et al., 1992); serotonin metabolism in normal aging brain is not impaired in contrast to AD (Yates et al., 1990).

For the cholinergic forebrain system controversial data on effects of aging have been reported. While McGeer et al. (1984) estimated a 70% loss of large cholinergic neurons of the nucleus basalis of Meynert (NBM)

Table 6. Neuronal Loss in Nucleus Basalis of Meynert in Aging and Alzheimer's Disease

	Neuron	Reduction	Density	
Condition	Region L	% Max	% Mean	Reference[a]
Normal	Ch 4	0		Chui et al., 1983
Aging	Ch 4			McGeer et al., 1984
(0–90 yrs)	Ch 4a		36	Mann et al., 1984
	Ch 4			Loew-Hummel et al., 1987
	Ch 1–4			Bigl et al., 1987
	Ch 1–4	0		Vogels et al., 1990
AD/DAT	Ch 4	90		Whitehouse, 1982
				Candy et al., 1983
		ChAT	18	Pearson et al., 1985
		AChE	14	
		62		Wilcock & Esiri, 1982
	Ch 1,2			Gertz et al., 1987
	Ch 2		39–70	Etienne et al., 1986
	Ch 4i		41–87	Doucette et al., 1986
	Ch 4		25–29	Rinne et al., 1987
	Ch 4p		57–89	Wilcock et al., 1988
	Ch 4	(large)		Allen et al., 1988
	Ch 4	0 54–73	22–76	Jellinger, 1989
	Ch 1,2			Vogels et al., 1990
	Total			
	Ch4	(58% large)		
	Ch 4p	(70% large)		
	Ch 4	53.3		Kobayashi et al., 1991

[a] See Jellinger (1991).

with age-related decrease of cortical cholin-acetyl transferase (CAT) activity, more recent automatized morphometric studies did not show any significant decrease in neuron numbers or density for the Chl to Ch4 sectors of NBM (Vogels et al., 1990; see Table 6), with no significant age-dependent variations in cortical CAT activity (Davies et al., 1988) or density of cholinergic fibers (Geula and Mesulam, 1989). However, a number of investigators have reported decreases with age in cortical CAT activity in individuals with preserved mental status (McGeer et al., 1984; Katzman et al., 1988).

In AD, magnocellular NBM cell loss ranges from 15% to 70% with significant depletion of large cholinergic neurons in the Ch4 region (Table

6). It correlates significantly with neuronal loss, density of AD lesions, and activity of presynaptic cholinergic marker enzymes and nicotinic receptors in the temporal target areas (Perry et al., 1989), suggesting retro- or anterograde degeneration of the NBM neurons (Allen et al., 1988). Additional shrinkage of large neurons suggests reduced enzyme synthesis or failure of transport as a result of neurotrophic factor dysfunctions (Vogels et al., 1990). Degeneration of the ascending innominato-cortical cholinergic system is associated with a severe decrease of cholinergic innervation of the cortex and hippocampus (Allen et al., 1988, 1990; Ransmayer et al., 1989) contributing to cognitive impairment in AD.

While the histaminic neurons in the hypothalamus are frequently involved by neurofibrillary tangles (NFTs) in AD (Airaksinen et al., 1991; Swaab et al., 1992), in the supraoptic and paraventricular nuclei of the hypothalamus, no significant differences in volume and total neuron numbers are seen between young and aged subjects of AD cases and controls, indicating that these nuclei are spared from degenerative changes in both senescence and AD (Goudsmith et al., 1990).

Dendritic and Synaptic Changes

Regional differences in the age-related changes of dendrites exist similar to those for neuronal loss. Continued dendritic regression with atrophy of dendritic trees is seen in aging frontal and temporal cortex (Scheibel, 1979; Nakamura et al., 1985), whereas in the parahippocampal cortex dendritic growth as a compensatory response to loss of neurons has been observed (Coleman and Flood, 1987). The progressive loss of apical dendrites and of postsynaptic dendritic spikes will cause partial deafferention of neurons (Haug, 1985). A significant age-related loss of synapses has been observed within the entire frontal cortex (Gibson, 1983) and in the precentral cortex with no changes in postcentral cortex (Adams, 1987).

Whereas in the motor association area 6 there is no age-related change in the density of synapses, the cortical areas 7 and 11 show a significant decrease in synapse density with constant length of synaptic densities (Haug, 1991). By contrast, in AD, cortical synaptic loss, as evaluated by direct counting at the ultrastructural level (Scheff et al., 1990; DeKosky and Scheff, 1990), by immunohistochemistry for synaptophysin (Masliah et al., 1991a; Terry et al., 1991) or both biochemistry and immunocytochemistry for synaptic markers (Lassmann et al., 1992),

Table 7. Correlation between Neurochemical Analysis of Synaptic and Large Dense-Core Vesicle Markers, Psychostatus (GDS), and Morphologic AD Markers in Temporal Cortex

Marker	GDS (n=14)	NFT (n=13)	SP (n=13)
Chromogranin A	0.0557		0.5703
	0.019	N.S.	0.021
Segretogranin II	-0.4730		
(S-II)	0.044	N.S.	N.S.
Synaptophysin	-0.6443	-0.748	-0.647
	0.006	0.001	0.008
SV 2 (synaptic marker)	-0.537	-0.802	-0.694
	0.024	0.000	0.004
p 65 (synaptic marker)	-0.674	-0.832	-0.714
	0.004	0.000	0.003
Chromogranin A/S. II	0.624	0.578	0.623
	0.009	0.019	0.012
Chromogranin A/SV 2	0.674	0.787	0.766
	0.004	0.001	0.001
Chromogranin A/p65	0.715	0.810	0.741
	0.002	0.000	0.002

First line: correlation coefficient; second line: significance level.
N.S.= not significant.
From Lassmann et al. (1992).

reveals a powerful correlation with cognitive impairment, biochemical changes, and the density of morphologic AD markers (Terry et al., 1991; Lassmann et al., 1992; Honer et al., 1992; Table 7).

Lipofuscin Deposition in Neurons

Accumulation of lipofuscin in neurons increasing with age in many regions of the brain with variable intensity is believed to result from defecits of lysosomal metabolism (Wisniewski and Wen, 1988). In general, it does not cause severe functional impairment, although massive deposition of lipofuscin is associated with decreased protein synthesis finally inducing neuronal dysfunction (Sohal and Wolfe, 1986). Immunoreactivity of lipofuscin with antibodies against β-amyloid protein (Bancher et al., 1989a) suggests specific disorders of neuronal degradation processes (Jellinger, 1990).

Alzheimer's Disease Markers and Aging Brain

The anatomical hallmarks of AD are focal extracellular deposition of βA4 amyloid protein (βAP) forming part of senile plaques (SP) and in cerebral vasculature, and an accumulation of cytoskeletal filaments in neuronal perikarya and dendrites producing NFTs and neuropil threads. Both types of lesion show various stages of development and may occur in variable intensity and distribution in brains of mentally intact elderly individuals and patients with AD.

Senile plaques show several stages of development (Delaere et al., 1991):

1. Stellate deposits, "early" or "very primitive" SP (Probst et al., 1991), consist of small granules with fibrillary processes showing strong βAP immunoreactivity with no central core or degenerating neurites (Davies et al., 1988; Ikeda et al., 1989; Ogomori et al., 1989; Yamaguchi et al., 1991).
2. Diffuse βAP deposits, preamyloid or "preplaques," or diffuse SP (Tagliavini et al., 1988; Wisniewski et al., 1989; Yamaguchi et al., 1991), show discontinuous amorphous βAP immunoreactive (IR) deposits along neurites and cell processes (Fig. 1A) with punctuate synaptophysin and chromogranin A(CA) IR (Brion et al., 1991; Probst et al., 1991, Masliah et al., 1991a; Lassmann et al., 1992) and presence of vesicles (Powers and Skeen, 1988; Yamaguchi et al., 1991, Masliah et al., 1991a). βAP is a 4.2-kDa polypeptide, a cleavage product of a larger amyloid precursor protein (APP), a membrane glycoprotein containing large extracellular domains and shorter intracytoplasmic regions encoded by a gene located on chromosome 21 (Selkoe, 1991; Holtzman and Mobley, 1991; Arai et al., 1991; Schubert et al., 1991). APP in brain is produced mainly by neurons (Card et al., 1988; Shivers et al., 1988; Arai et al., 1991), although non-neuronal components of brain tissue as well as non-neuronal tissues also express this protein (see Arai et al., 1991; Mann et al., 1992). Recent studies indicate that APP is preferentially found at synaptic sites where it is colocalized with SYN (Schubert et al., 1991). APP is further found on vesicular elements of neuronal processes, dendrites, and axons (Allsop et al., 1988; Cras et al., 1990; El Hachimi et al., 1991; Probst et al., 1991). Within early diffuse SP and βAP deposits, neurons and dilated SYN-IR presynaptic termi-

nals and/or neuronal vesicles are consistently identified (Powers and Skeen, 1988; Pappola et al., 1991; Masliah et al., 1991a,b). These data suggest that βAP deposits begin on a neuronal level, probably at the presynaptic level (Bugiani et al., 1990; Masliah et al., 1991; Schubert et al., 1991). Changes in the transcription and post-translational processing of βAPs produced in the brain are likely to play a role in the accumulation of βAP in normal aging, Down's syndrome and AD (Selkoe, 1991; Arai et al., 1991; Hardy and Higgins, 1992; Joachim and Selkoe, 1992). In advanced stages of diffuse SP small degenerating SYN- and CA-IR neurites, astroglial processes, and neurons are associated with amyloid fibers and electron-dense materials forming transitional stages to "primitive" SP (Yamaguchi et al., 1991).

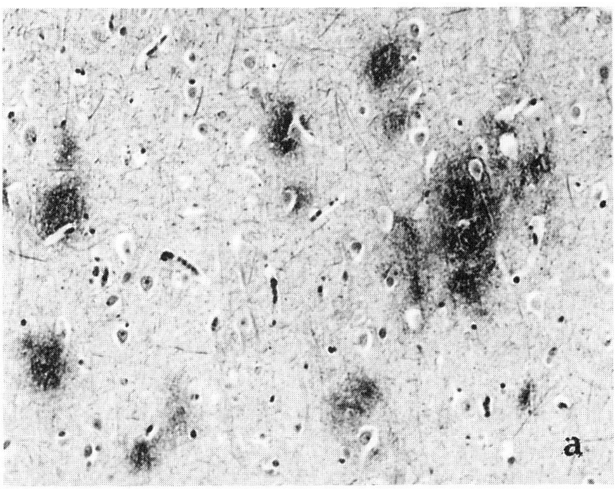

Figure 1A Diffuse β-amyloid protein deposits in frontal cortex stained with monoclonal antibody 4G8 $\times$ 50.

3. "Primitive" SPs contain round-shaped, well demarcated amyloid deposits, between degenerating SYN-IR and CA-IR neurites and reactive astrocytes and microglia (Ikeda et al., 1989; Fig. 1B).
4. Classic or mature "core" (neuritic) SPs show a central amyloid core with a βAP crown surrounded by dystrophic or regenerating neurites and microglia (Fig. 1C).
5. Compact or burnt-out SPs contain only an amyloid core without a βAP crown.

βAP deposits are found in brains of both mentally intact aged and AD individuals with the same and relatively uniform patterns of distribution in both groups (Ogomori et al., 1989; Tagliavini et al., 1990; Cras et al., 1991). βAP and APP deposits also have been found in rodents and mammals (e.g., dogs, bears, horses and nonhuman primates) (cf. Selkoe et al., 1987; Wisniewski et al., 1990; Ishihara et al., 1991; Podlisky et al., 1991). Demonstration of homology of the APP and βAP in monkey and human brain (Podlisky et al., 1991) suggests that βAP deposition may be a nonspecific property of aging. Amorphous deposits of βAP are found in the brains of young-adult Down's syndrome patients before the other cytopathologic features of AD develop (Allsop et al., 1988; Ikeda et al., 1989; Mann and Esiri, 1989, Giaccone et al., 1989) and in nondemented aged individuals (Ulrich, 1985; Davies et al., 1988; Duyckaerts et al., 1988; Mann et al., 1990; McKee et al., 1991). Recent observations of βAP deposits in brains of transgenic rodents overexpressing a C-terminal fragment of human APP supports early transcription and post-translational processing of APP that precedes βAP deposition (Quon et al., 1991; Sandhu et al., 1991; Wirak et al., 1991; Kawahata et al., 1991). *NFTs*, another morphologic hallmark of AD, are abnormal deposits of neuronal cytoskeleton formed by paired helical and straight filaments (Wisniewski and Wen, 1985) containing phosphorylated microtubule-associated tau proteins resulting from pathologic phosphorylation of tau followed by progressive ubiquitination resulting from disturbed protein decomposition (Bancher et al., 1989b, 1991).

The origin and relationship between both types of lesion are unknown. Brain amyloidosis is not necessarily accompanied by NFT formation, and, although βAP deposition is relatively uniform throughout neocortex and hippocampus in aged persons (Cairns et al., 1991), the development and distribution of SPs and NFTs are independent from each other (Zubenko et al., 1989; Probst et al., 1989; Armstrong, 1991; Armstrong et al., 1991; Dickson et al., 1992).

Whereas neocortical NFTs and neuritic SP are usually absent in most mentally intact aged persons, βAP deposits increase with age and may affect up to 80% and more in their 80s (Davies et al., 1988) but may not be an invariable accompaniment of aging. The percentage of nondemented elderly persons showing no neocortical SP and/or βAP deposits demonstrable by any of the employed histologic staining methods ranges from 10% to 46% (Terry et al., 1987b; Crystal et al., 1988; Katzman et al., 1988, Davies et al., 1988; Ikeda et al., 1989; Mann et al., 1990; Braak and

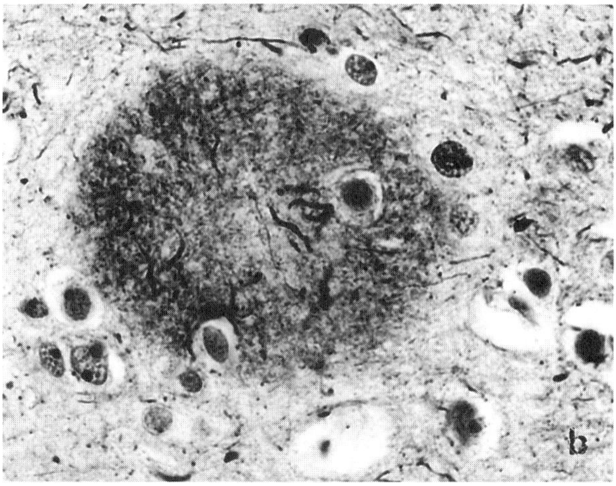

Figure 1B Primitive plaque without central amyloid core or dystrophic neurites. Modified Bielschowsky silver stain ×200.

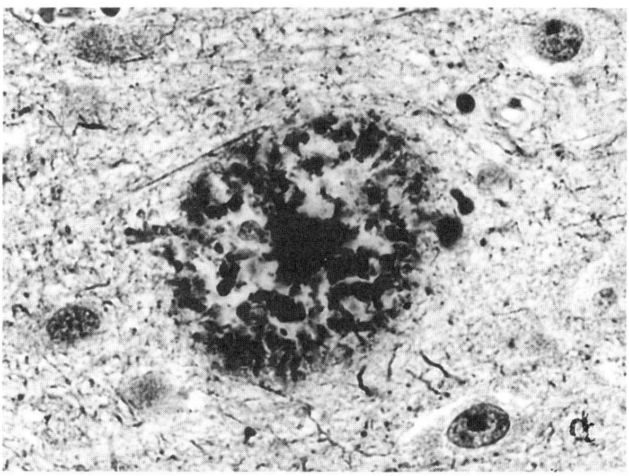

Figure 1C Classical plaque with central amyloid core and peripheral crown of dystrophic neurites. Modified Bielschowsky stain ×200.

Braak, 1991; McKee et al., 1991) and, in single cohorts, even amounts up to 50% (Morris et al., 1991) or 69% (Price et al., 1991). These data suggest that "normal" aging may occur without any AD lesions in the brain, and neocortical SP and NFT may be absent in an age range of 75 to 90 years (Morris et al., 1991). While neocortical SPs are rare in mentally intact persons under the age of 60 years, the incidence of neocortical βAP deposits or SPs, largely of the diffuse type, in nondemented persons older than 65 years ranges from 33% to 80% (Table 8; Fig. 2) and, in general, shows a linear age-dependent increase. In the hippocampal region of the mentally intact seniors, the incidence of SPs ranges from 0% to 33% in the 6th decade to 94 up to 100% in the 10th decade (Jellinger, 1989), although they may be absent in single nondemented centenaries (Table 9). The density of SPs is correlated to the reduction of neurons in the hippocampus (Ball, 1988), but not in neocortex (Miyoshi and Sato, 1991). While Gentleman et al. (1991) found significant correlations between mental test scores and the amount of βAP in entorhinal cortex and hippocampal subfields in both AD and aged controls, other studies of the temporal isocortex in persons with graded psychostatus revealed significant correlations between mental impairment and density only of neuritic changes (neuritic SPs and NFTs) but not of diffuse βAP deposits (Delaere et al., 1991; Morris et al., 1991; McKee et al., 1991; Arriagada et al., 1992b). On the other hand, mentally intact elderly persons and those with very mild dementia may harbor large amounts of neocortical βAP deposits and SPs, particularly of the diffuse type (see Table 8) that, according to recent immunohistochemical studies, do not accentuate synaptic loss (Masliah et al., 1990). These and other data suggest that (a) βAP deposits and diffuse SP in the absence of neuritic pathology appear to be part of normal aging with no clinical significance, and (b) that these changes may exist without development of dementia. Hence, βAP deposits appear to be necessary but not a sufficient condition for inducing neuritic components and thus may not be, per se, markers of incipient AD related to intellectual impairment (Mann et al., 1990; McKee et al., 1991; Jellinger et al., 1991; Dickson et al., 1992).

Whereas neuritic lesions (neuritic SPs and NFTs) in neocortex may be absent in 8% to 36% of demented individuals over the age of 75 years, most of them showing large numbers of NFTs in hippocampus (Terry et al., 1987b; Joachim et al., 1988; Morris et al., 1991; McKee et al., 1991), the intensity of neuritic pathology usually shows significant correlations with both mental impairment (Crystal et al., 1988; Delaere et al., 1988, 1991; Morris et al., 1991; McKee et al., 1991; Jellinger et al., 1991; Fischer et al., 1991; Arriagada et al., 1992b) and biochemical changes, such as cortical CAT activity and synaptic markers (Bowen et al., 1985; Terry et al., 1987b, 1991; Mountjoy, 1988; Esiri et al., 1990; Lassman et al., 1992). On the other hand, the presence of neuritic SPs and NFTs is not necessarily associated with intellectual impairment: whereas in some case series neocortical NFTs were not observed in mentally intact persons

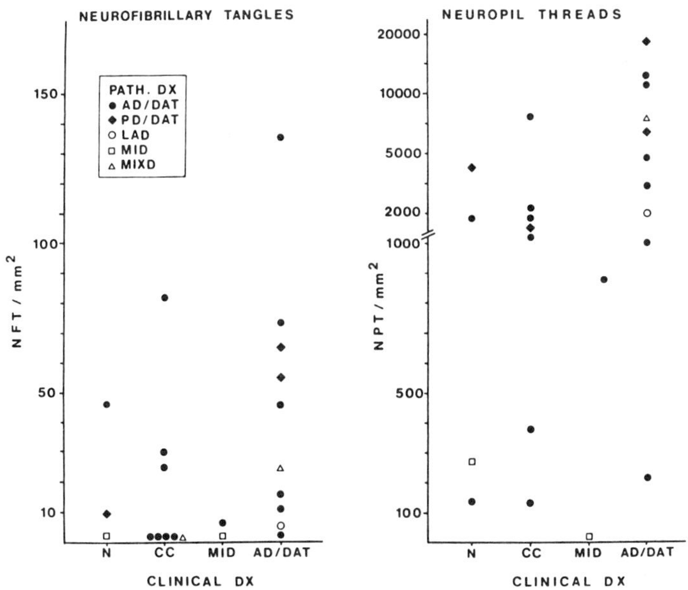

Figure 2 Total plaque counts in temporal and parahippocampal cortex of 27 elderly patients. Three or four nondemented patients (*N*) aged 67–96 (mean 80 yrs) with MMS of 28 to 30 showed 8 to 32 senile plaques/mm^2; six of 8 patients with cognitive changes but no dementia (MMS 22–27) and all but one severely demented AD cases showed large numbers of plaques. DX, diagnosis; AD/DAT, Alzheimer's disease; LAD, limbic type of AD; MID, multiinfarct dementia; MIXD, mixed type dementia.

over age 60 years (Crystal et al., 1988; Jellinger et al., 1991; Morris et al., 1991; Delaere et al., 1991), Tomlinson and Henderson (1976) found occasional neocortical NFTs in 10% to 20% of the nondemented elderly. In other cohorts, occasional neuritic SPs and/or NFTs in temporal neocortex were observed in 20% (McKee et al., 1991; Lassmann et al., 1992) to around 50% (Arriagada and Hyman, 1990; Mann et al., 1990; Price et al., 1991), whereas Price et al. (1991) saw PHF- and Alz-50–positive neurons in 41% and 92%, respectively.

Much more intensive neuritic changes, particularly NFTs, in the limbic system are frequently observed in nondemented aged persons; their incidence ranging from 20% to 92% (Table 10; Fig. 3). Universal and early NFTs in layer II of entorhinal cortex are present in Down's syndrome (Mann and Esiri, 1989; Hyman and Mann, 1991), AD (Hyman et al., 1986, 1988; Braak and Braak, 1987, 1991; Price et al., 1991; Jellinger et al., 1991), and in normal aging (Ulrich, 1985; Arriagada et al., 1992a; Dickson et al., 1992; Price et al., 1991; Morris et al., 1991;

Table 8. Neocortical βAP Deposits (Senile Plaques) in Nondemented Elderly Persons

Reference	Percent	Type of Neocortical Lesion
Tomlinson (1980)	9(5/56)	Excess neocortical SPs
Davies et al. (1988)	55(over age 65)	βAP deposits
Crystal et al. (1988)	75(3/4) (mean age 86)	Primitive SPs
	25(1/4)	Numerous SPs frontal cortex
Katzman et al. (1988)	3(10/30)	Numerous neocortical SPs
Delaere et al. (1989)	75(4/5)	Numerous neocortical SPs
Mann et al. (1990)	75(12/15) (over age 65)	βAP deposits
	53(8/15) (over age 60)	βAP+neuritic SPs
McKee et al. (1991)	54(12/22) (mean age 74)	βAP in isocortex
Price et al. (1991)	30(4/13) (over age 54)	βAP in isocortex
Braak & Braak (1991)	64(28/44) (mean age 72)	Same
Morris et al. (1991)	50(2/4) (age 76–89)	Same
Jellinger et al. (1991)	75(3/4) (age 66–97)	Same
Lassmann et al. (1992)	60(3/5) (age 66–96)	Same

McKee et al., 1991; Braak and Braak, 1991; Jellinger et al., 1991). This strongly supports the hypothesis that changes in the neurons of layer II of entorhinal cortex are one of the earliest manifestations of aging and/or of AD pathology. On the other hand, Brady and Mufson (1991), in nondemented aged controls aged 66 to 76 years, found Alz-50 immunoreactive neurons most frequently in the CA-2 sector of hippocampus and only occasionally in layer II of entorhinal cortex. Although the mesolimbic cortex and hippocampus are thought to be central in the integration of emotional responses, behavior, and intellect, presence of neuritic SPs and NFTs in entorhinal cortex and/or hippocampus formation do not appear to be a reliable predictor of dementia, as a fairly large number of mentally intact elderly show neuritic lesions in these regions and cannot be clearly distinguished from demented persons (Price et al., 1991; Dickson et al., 1992).

Threshholds Between Aging and AD

The postmortem diagnosis of AD currently based on aged-adjusted quantitative histologic criteria (Khachaturian, 1985; Tierney et al., 1988) and the use of reference staining methods to assess AD markers (Lamy et

Table 9. Alzheimer Lesions in Hippocampus of Nondemented Elderly

Reference	N	Age Range	SP (%)	NFT (%)
Hirano & Zimmerman (1962)	30	70–90 A	50	77.8
Tomlinson (1968)	28	68–92 ($\tilde{x}$ 75)	78	62
Dayan (1970)	50	6th decade	0	0
		7th decade	21.4	0
Wilcock & Esiri (1982)	14	65–102 ($\tilde{x}$ 80)	64.3	42.8
Ulrich (1982)	100	6th decade	14	18
Ulrich (1984)	51	6th decade	25.5	18
Jellinger (1986)	250	7th decade (57)	18.5	17
		8th decade(116)	88.8	72.5
		9th decade(65)	93.8	90.8
		10th decade (12)	100	100
Alafuzoff et al. (1987)	15	46–91 ($\tilde{x}$ 74)	68	42
Mann et al. (1987)	28	6th decade	33	90
		7th decade	100	90
		8th decade	90	90
Crystall et al. (1988)	4	mean 86	25	33
Mann et al. (1990)	15	60–84	80	73
Price et al. (1991)	12	54–83 Ca 1	7.5	66
		amygdala	7.5	46
	8	73–88 CA 1	12	100
		amygdala	12	88
McKee et al. (1991)	22	66-99 ($\tilde{x}$ 75)	54	64
Unger et al. (1992)	7	51–89 amygdala	0	few

[a] See Jellinger, (1989).

al., 1989; Duckyaerts et al., 1990; Vallet et al., 1992) not necessarily imply uniform results and leave room for controversy in the *interrater* reliability (Paulus et al., 1992). Tolerance of the histologic AD markers with senescence and difficulties in the determination of intellectual integrity further blur the distinction between normal aging and AD that will become a matter of quantitation (Xuereb et al., 1991). Therefore, it may be difficult (a) to differentiate with any certainty, autopsy cases of early ot mild AD from mentally intact aged subjects, and (b) to determine whether the density and distribution of βAP, SP, and NFT in nondemented elderly are related to the normal aging process or to incipient AD. These factors have suggested for some that the two conditions form a clinicopathologic spectrum (Gottfries et al., 1990), and that a

Table 10. Limbic NFTs in Nondemented Elderly Persons

Reference	Percent incidence	Age Range	Region
Arriagada-Hyman (1990)	100(15/15)	58–85 (×70.4)	Entorhinal II/IV
	66(10/15)		CA 1
	26(4/15)	Isolated NFT	Entorhinal II/IV
Mann et al. (1990)	33(1/3)	(Under age 60)	Hippocampus
Jellinger et al. (1991)	50(2/4)	66–96 ($\tilde{x}$80)	Entorhinal
Morris et al. (1991)	75(3/4)	73–87	Entorhinal
McKee et al. (1991)	64(14/22) (9 with, 5 without βAP in isocortex)	66–95 (x71.4)	Entorhinal
Braak & Braak (1991)	75(33/44)	($\tilde{x}$ 72)	Entorhinal
Price et al. (1991)	92(12/13)	54–89	Entorhinal
	100(5/5)	Under age 65	A-50 IR entorhinal
	40(2/5)		A-50 IR CA 1
	75(4/5)		NFT entorhinal
Lassmann et al. (1982)	25(1/5)	66–97	PHF parahippocampal gyrus

study of mentally intact people may include clinically undetected (pre- or subclinical) AD cases (Mann et al., 1990; Morris et al., 1991; Price et al., 1991). Additional evidence supporting a long (preclinical) period of illness with progression of anatomic lesions in nondemented persons is the appearance of βAP deposits in young Downs syndrome patients, suggesting that deposition of APP or βAP begins around 30 years before manifestation of cognitive deficits (Hyman and Mann, 1991; Murphy and Ellis, 1991).

Another unresolved problem is the relationship between AP deposits currently considered as the initial or central event in the etiology of AD (Selkoe, 1991; Hardy and Allsop, 1991; Schubert et al., 1991; Hyman and Tanzi, 1992; Joachim and Selkoe, 1992) and other cytopathologic AD changes. The fact that APP and βAP deposits are frequently found in the brains of nondemented aged persons and of young Down syndrome patients in the absence of neuritic AD lesions has led to the assumption that deposition of βAP precedes the development of neuritic AD pathology (Giaccone et al., 1988). The first affected areas are the entorhinal cortex, hippocampus, and amygdala. Since neocortical βAP deposits and diffuse SP are seen in both mentally intact elderly and those with very

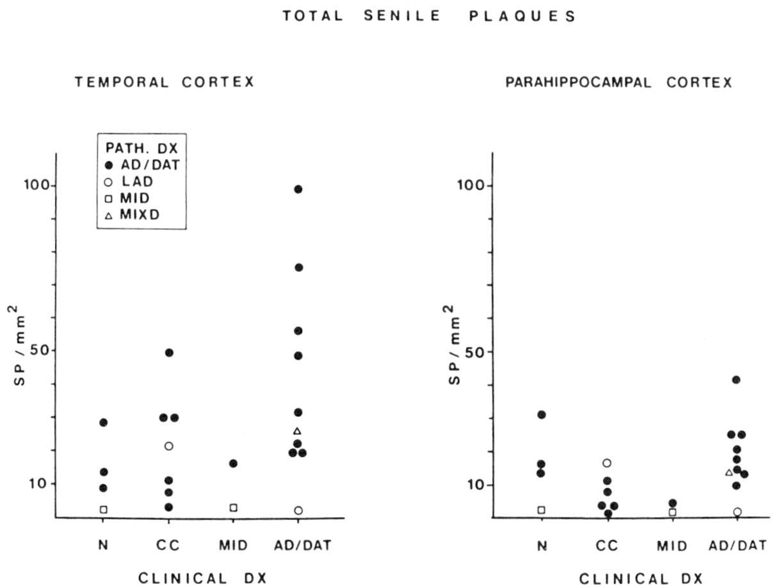

Figure 3 Number of neurofibrillary tangles and neuropil threads/mm^2 in entorhinal cortex in a series of 27 elderly patients with graded psychostatus (same as in Fig. 2). Two of three nondemented controls had 10 to 48 NFTs/mm^2, and two of the four nondemented controls had large numbers of neuropil threads in this region. Legend as in Fig. 2.

mild dementia, it can be argued that these lesions represent the earliest signs of incipient AD and are indicative of a disease process related to dementia. This is supported by experimental evidence of neurotoxic and immunologic effects of injected AD-βAP in rat brain (Frautchy et al., 1991). On the other hand, in a number of nondemented elderly, cortical NFTs precede the deposition of βAP, particularly in the limbic system, suggesting that the early appearance of NFTs may be related to aging more than to AD (Braak and Braak, 1991; Price et al., 1991; Arriagada et al., 1992). This is supported by the fact that widespread neuritic AD pathology may be unrelated to βAP deposits (Tabaton et al., 1988; Probst et al., 1989; Price et al., 1991; Armstrong, 1991; Braak and Braak, 1991; Jellinger, 1991b; Dickson et al., 1992). The existence of βAP deposits, therefore, may not be considered a prerequisite of neuritic AD pathology and synaptic loss, both being strongly related to dementia. In conclusion, recent data suggest that various factors contribute to the development of mental dysfunction in aging and AD, the definite distinction of which may be difficult. At present, the borders between "normative" brain aging and AD are not clear, and the etiology of changes in the aging brain awaits further elucidation. Further longitudinal research with quantitative studies are morphologic, biochemical, and genetic markers in the brains of aged persons with graded mental status will be necessary for a better understanding of the relationship and limits between normal aging and its threshold levels in the evolution of dementia.

Summary

"Normative" brain aging is featured by reduction in volume and weight of the brain and loss and reduction in size of large neurons in the frontal and temporal cortex with no or only few changes in the hippocampus and most of the subcortical projection nuclei, which may be caused by the absence or only mild occurence of cytoskeletal lesions. On the other hand, both lipofuscin in neuronal perikarya and β-amyloid deposits in neocortex and hippocampus occur with increasing age, but their implications for mental impairment are not clear, since both changes and rare neurofibrillary tangles in hippocampal formation may occur in both mentally intact elderly and in AD patients. Based on the data from recent literature and personal experience, it appears difficult to distinguish between the aging brain and AD, and further studies will be necessary to determine the thresholds between normative aging and Alzheimer's disease.

References

Adams I (1987): Comparison of synaptic changes in the precentral and postcentral cerebral cortex of aging humans. A quantitative ultra-structural study. *Neurobiol Aging* 8:203–212.

Aletrino MA, Vogels OJM, Van Domburg PHMF, Ten Donkelaar HJ (1992): Cell loss in the nucleus raphes dorsalis in Alzheimer's disease. *Neurobiol Againg* 13:461–468.

Airaksinen MS, Patau A, Paljarvi L, et al. (1991): Histamine neurons in human hypothalamus. Anatomy in normal and Alzheimer diseased brains. *Neuroscience* 44:465–481.

Allen SJ, Dawbarn D, MacGowan SH, Wilcock GK, Treanor JJS, Moss TH (1990): Quantitative morphometric analysis of basal forebrain neurons expressing β-nerve growth factor receptors in normal and Alzheimer's disease brains. *Dementia* 1:125–137.

Allen SJ, Dawbarn D, Wilcock CK (1988): Morphometric immunochemical analysis of neurons in the nucleus basalis of Meynert in Alzheimer disease. *Brain Res* 454:275–281.

Allsop D, Haga SI, Haga C, Ikeda SI, Mann DMA, Ishii T (1989): Early senile plaques in Down's Syndrome brains show a close relationship with cell bodies of neurons. *Neuropathol Appl Neurobiol* 15:531–542.

Arai H, Lee VM, Messinger ML, Greenberg BD, Lowery DE, Trojanowski JO (1991): Expression patterns of β-amyloid precursor protein (β-APP) in neural and nonneural human tissues from Alzheimer's disease and control subjects. *Ann Neurol* 30:686–693.

Armstrong RA (1991): The relationship between senile plaques, neurofibrillary tangles and amyloid (A4) protein deposits in Alzheimer's disease. *Neurosci Res Commun* 9:91–97.

Armstrong RA, Myers D, Smith CUM (1991): Alzheimer's disease: the relationship between the density of senile plaques, neurofibrillary tangles and A4 protein in human patients. *Neurosci Lett* 123:141–143.

Arriagada PV, Marzloff K, Hyman BT (1992a): Distribution of Alzheimer-type pathologic changes in nondemented elderly individuals matches the pattern in Alzheimer's disease. *Neurology* 42:1681–1688.

Arriagada PV, Growdon JH, Hedley-White ET, Hyman BT (1992b): Neurofibrillary tangles but not senile plaques parallel duration and severity of Alzheimer's disease. *Neurology* 42:631–639.

Arriagada P, Hyman RT (1990): Topographic distribution of Alzheimer neuronal changes in normal aging brains. An immunohistochemical study with Alz-50. *J Neuropathol Exp Neurol* 49:336.

Ball MJ (1988): Hippocampal histopathology — a critical substrate of dementia of the Alzheimer type. *Interdiscipl Topics Gerontol* 25:16–37.

Bancher C, Grundke-Iqbal I, Iqbal K, Fried VA, Smith HT, Wisniewski HM (1991): Abnormal phosphorylation of tau precedes ubiquitination of neurofibrillary pathology of Alzheimer disease. *Brain Res* 539:11–18.

Bancher C, Grundke-Iqbal I, Iqbal K, Kim KS, Wisniewski HM (1989a): Immunoreactivity of neuronal lipofuscin with monoclonal antibodies to the amyloid β-protein. *Neurobiol Aging* 10:125–132.

Bancher C, Brunner C, Lassmann H, Budka H, Jellinger K, et al. (1989b): Accumulation of abnormally phosphorylated tau precedes the formation of neurofibrillary tangles in Alzheimer's disease. *Brain Res* 477:90–99.

Bhatnagar K, Kennedy RC, Baron G, Greenberg RA (1987): Number of mitral cells and the bulb volume in the aging human olfactory bulb: a quantitative morphological study. *Anat Rec* 218:73–87.

Bowen DM, Davison AN, Francis PT, Palmer AM, Pearce BR (1985): Neurotransmitter and metabolic dysfunctions in Alzheimer's dementia. In: *Modern Approaches to the Dementias*, Rose FC., ed. Basel: S. Karger.

Braak H, Braak E (1991): Neuropathological stageing of Alzheimer-related changes. *Acta Neuropathol* 82:239–259.

Braak H, Braak E (1990): Neurofibrillary changes confined to the entorhinal region and an abundance of cortical amyloid in cases of presenile and senile dementia. *Acta Neuropathol* 80:479–486.

Brady DB, Mufson EJ (1991): Alz-50 immunoreactive neuropil differentiates hippocampal complex subfields in Alzheimer's disease. *J Comp Neurol* 305:489–507.

Brion JP, Couck AM, Bruce M, Anderton B, Flament-Durand J (1991): Synaptophysin and chromograin A immunoreactivities in senile plaques of Alzheimer's disease. *Brain Res* 539:143–150.

Brody H (1955): Organization of the cerebral cortex. III. A study of aging in the human cerebral cortex. *J Comp Neurol* 102:511–556.

Brown MW, Cassell MD (1980): Estimates of the number of neurones in the human hippocampus. *J Physiol* 301:58–59P.

Bugiani O, Giaccone G, Verga L, Pollo B, Ghetti B, Frangione B, Tagliavani F (1990): Alzheimer patients and Down patients: abnormal presynaptic terminals are related to cerebral preamyloid deposits. *Neurosci Lett* 119:56–59.

Cairns NJ, Chadwick A, Luthert PJ, Lantos PJ (1991): β-amyloid protein load is relatively uniform throughout neocortex and hippcampus in elderly Alzheimer disease patients. *Neurosci Lett* 129:115–H118.

Card JP, Meade RP, Davis LG (1988): Immunocytochemical localization of the precursor protein for β-amyloid in the rat central nervous system. *Neuron* 1:835–846.

Chan-Palay V, Asan E (1989): Quantitation of catecholamine neurons in the locus coeruleus in human brains of normal young and older adults and in depression. *J Comp Neurol* 287:357–372.

Coleman PD, Flood DG (1987): Neuron numbers and dendritic extent in normal aging and Alzheimer's disease. *Neurobiol Aging* 8:521–545.

Colon EJ (1972): The elderly brain. A quantitative analysis of the cerebral cortex of two cases. *Psychiatr Neurol Neurochir* 75:261–270.

Cras P, Kawai M, Siedlak S, et al. (1990): Neuronal and microglial involvement in β-amyloid protein deposition in Alzheimer's disease. *Am J Pathol* 137:241–246.

Crystal H, Dickson D, Fuld P, et al. (1988): Clinico-pathologic studies in dementia: Nondemented subjects with pathologically confirmed Alzheimer's disease. *Neurology* 38:1682–1687.

Davies DC, Horwood N, Isaacs SL, Mann DMA (1992): The effect of age and Alzheimer's disease on pyramidal neuron density in the individual fields of the hippocampal formation. *Acta Neuropathol* 83.

Davies L, Wolksa B, Hilbich C, et al. (1988): A4 amyloid protein deposition and the diagnosis of Alzheimer's disease. Neurology 38:1688–1693.

DeKosky ST, Scheff SW (1990): Synapse loss in frontal cortex biopsies in Alzheimer's disease: correlation with cognitive severity. *Ann Neurol* 27:457–464.

Delaere P, Duyckaerts C, Brion PJ, Poulain V, Hauw JJ (1989): Tau, paired helical filaments and amyloid in the neocortex: a morphometric study of 15 cases with graded intellectual status in aging and senile dementia of Alzheimer type. *Acta Neuropathol* 77:645–653.

Delaere P, Duyckaerts C, He Y, Piette F, Hauw JJ (1991): Subtypes and different laminar distributions of βA4 deposits in Alzheimer's disease: relationship with the intellectual status of 26 cases. *Acta Neuropathol* 81:328–335.

Dickson DW, Crystal HA, Mattiace LA, Masur DM, Blau AD, Davies P, Yen SH, Aronson MK (1992): Identification of normal and pathological aging in prospectively studied nondemented elderly humans. *Neurobiol Aging* 13:179–189.

Duyckaerts C, Delaere P, Hauw JJ, et al. (1990): Rating of the lesions in senile dementia of the Alzheimer type: concordance between laboratories. A European multicenter study under the auspices of Eurage. *J Neurol Sci* 97:295–323.

Duyckaerts C, Delaere P, Poulain V, Brion JP, Hauw JJ (1988): Does Amyloid precede paired helical filaments in the senile plaques? *Neurosci Lett* 91:354–359.

El Hachimi KH, Verga L, Giaccone G, Tagliavini F, et al. (1991): Relationship between non-fibrillary amyloid precursors and cell processes in the cortical neuropil of Alzheimer patients. *Neurosci Lett* 129:119–122.

Esiri M, Pearson RCA, Steele JE, Bowen DB, Powell TPS (1990): A quantitative study of the neurofibrillary tangles and the choline acetyl transferase activity in the cerebral cortex and the amygdala in Alzheimer's disease. *J Neurol Neursurg Psychiatr* 53:161–165.

Fearnley JM, Lees AJ (1991): Ageing and Parkinson's disease. Substantia nigra regional selectivity. *Brain* 114:2283–2301.

Fischer P, Lassmann H, Jellinger K, et al. (1991): Die Demenz vom Alzheimer-Typ. Eine klinische Laengsschnittstudie mit quantitativer Neuropathologie. *Wien Med Wschr* 141:455–462.

Flood DG (1991): Region-specific stability of dendrite extent in normal human aging and regression in Alzheimer's disease. II. Subiculum. *Brain Res* 540:83–95.

Förstl H, Burns A, Cairns N, Luthert P, Lantos P, Levy R (1992): Organische Grundlagen depressiver Symptome bei der Alzheimer-Demenz. Ergebnisse einer prospektiven Untersuchung; Literaturüberisicht. *Nervenarzt* 63:566–574.

Frautschy SA, Baird A, Cole GM (1991): Effects of injected Alzheimer β-amyloid cores in rat brain. *Proc Natl Acad Sci USA* 88:8362–8366.

Gentleman S, Perry RM, Roberts GW (1991): Correlations between β-amyloid protein (βAP) load and mental test scores in Alzheimer's disease. *Neuropathol Appl Neurobiol* 17:531.

German DC, Manaye KF, White CL III, Woodward DJ, McIntire DD, Smith WK, Kalarisa RN, Mann DMA (1992): Disease-specific patterns of locus coeruleus cell loss. *Ann Neurol* 32:667–676.

Geula C, Mesulam MM (1989): Cortical cholinergic fibers in aging and Alzheimer's disease. A morphometric study. *Neuroscience* 33:469–481.

Giaccone G, Tagliavini F, Linosli G et al. (1989): Down patients: extracellular preamyloid deposits precede neuritic degeneration and senile plaques. *Neurosci Lett* 97:232–238.

Gibson PH (1983): EM-study of the numbers of cortical synapses in the brains of aging people and people with Alzheimer-type dementia. *Acta Neuropathol* 62:127–133.

Gottfries CG, Blenow K, Regland B, et al. (1990): Alzheimer's disease - one, two or several? In: *Alzheimer's Disease*, Maurer K, et al., eds.: Springer-Verlag, Vienna.

Goudsmith E, Hofman MA, Fliers E, Swaab DF (1990): The supraoptic and paraventricular nuclei of the human hypothalamus in relation to sex, age, and Alzheimer's disease. *Neurobiol Aging* 11:529–536.

Halliday GM, McCann HL, Pamphlett R, Brooks WS, Creasey L (1992): Brainstem serotonin-synthesizing neurons in Alzheimer's disease: a clinico-pathological correlation. *Acta Neuropathol* 84:638–650.

Hanks SD, Flood D (1991): Region-specific stability of dendritic extent in normal human aging and regression in Alzheimer's disease. II. CA1 of hippocampus. *Brain Res* 540:63–82.

Hardy IA, Higgins GA (1992): Alzheimer's disease: The amyloid cascade hypothesis. *Science* 256:184–185.

Hardy J, Allsop D (1991): Amyloid deposition as the central event in the aetiology of Alzheimer's disease. *Trends Pharm Sci* 12:283.

Haug H (1985): Are neurons of the human cerebral cortex really lost during aging? In: *Senile Dementia of the Alzheimer type*, Traber J, Grispen WG, eds., Berlin: Springer-Verlag.

Haug H (1991): Morphometric evaluation of the aged human brain. Changes in the volume of various grays, in the neuronal composition and ultrastructure. *Neuropathology* 4(Suppl.):631–634.

Holtzman DM, Mobley WC (1991): Molecular studies in Alzheimer's disease. *Trends Biol Sci* 16:140–144.

Honer WG, Dickson DW, Gleeson J, Davies P (1992): Regional synaptic pathology in Alzheimer's disease. *Neurobiol Aging* 13:375–382.

Hornykiewicz O, Pifl C, Kish SJ, Shinnak K, Schingnitz G (1988): Biochemical changes in idiopathic Parkinson's disease, aging, and MPTP parkinsonism: similarities and differences. In: *Parkinsonism and Aging*, Calne DB, ed., New York: Raven Press.

Hyman BT, Tanzi RE (1992): Amyloid, dementia and Alzheimer's disease. *Curr Opin Neurol Neurosurg* 5:88–93.

Hyman BT, Mann DMA (1991): Alzheimer-type pathological changes in Down's syndrome individuals of various ages. In: *Alzheimer's Disease: Basic Mechanisms, Diagnosis and Therapeutic Strategies*, Iqbal K, et al., eds. Chichester: John Wiley & Sons.

Hyman BT, Van Hoesen GW, Wolozin BL, Davies P, Krommer LJ, Damasio AR (1988): Alz-50 antibody recognizes Alzheimer-related neuronal changes. *Ann Neurol* 23:371–379.

Hyman BT, Van Hoesen GW, Damasio AR (1986): Alzheimer's disease: glutamate depletion in the hippocampal perforant pathway. *Ann Neurol* 22:37–40.

Ikeda S, Yanagisawa N, Allsop D, Glenner GG (1990): Early senile plaques in Alzheimer's disease demonstrated by histochemistry, immunocytochemistry, and electron microscopy. *Hum Pathol* 21:1221-1226.

Ince P, Irving D, MacArthur F, Perry RH (1991): Quantitative neuropathological study of Alzheimer-type pathology in the hippocampus: comparison of senile dementia of Alzheimer type, senile dementia of Lewy body type, Parkinson's disease and non-demented elderly control patients. *J Neurol Sci* 106:142–152.

Ishihara T, Gondo T, Takahashi M, Uchino F, et al. (1991): Immunohistochemical and immunoelectron microscopical characterization of cerebrovascular and senile plaque amyloid in aged dogs' brains. *Brain Res* 548:196–205.

Jellinger K (1991a): Pathology of Parkinson's disease. Changes other than the nigrostriatal pathway. *Mol Chem Neuropathol* 14:153–197.

Jellinger K (1991b): Aging and Alzheimer disease. *Neurology* 41:1852.

Jellinger K, Braak H, Braak E, Fischer P (1991a): Alzheimer lesions in entorhinal region and isocortex in Parkinson's and Alzheimer's diseases. *Ann NY Acad Aci* 640:203–209.

Jellinger K, Bancher C, Fischer P, Lassmann H (1991b): Quantitative histopathologic validation of senile dementia of Alzheimer type. A prospective clinicopathologic study. *Eur J Gerontol* 1(3).

Jellinger K (1990): Morphology of Alzheimer's disease and related disorders. In: *Alzheimer's Disease: Epidemiology, Neuropathology, Neurochemistry, and Clinics*, Maurer K, et al., eds. Wien-New York: Springer Verlag.

Jellinger K (1989): Morphologie des alternden Gehirns und der (prä-)senilen Demezen. In: *Handbuch der Gerontologie, Band 5, Neurologie, Psychiatrie*, Platt D, Oesterreich K, eds. Stuttgart-New York: Gustav Fischer Verlag.

Joachim CL, Selkoe DJ (1992): The seminal role of β-amyloid in the pathogenesis of Alzheimer disease. *Alzheimer Disease and Assoc Disordr* 6:7–34.

Katzman R, Terry R, DeTeresa R, et al. (1988): Clinical, pathological and neurochemical changes in dementia. *Ann Neurol* 23:138–144.

Kawabata S, Higgins GA, Gordon JW (1991): Amyloid plaques, neurofibrillary tangles and neuronal loss in brains of transgenic mice overexpressing a C-terminal fragment of human amyloid precursor protein. *Nature* 54:476–478.

Kemper T (1984): Neuroanatomical and neuropathological changes in normal aging and dementia. In: *Clinical Neurology of Aging*, Albert ML, ed. New York: Oxford Univ. Press.

Keyer J De, Ebinger G, Vauquelin G (1990): Age-related changes in the human striatonigral dopaminergic system. *Ann Neurol* 27:157.

Khachaturian ZS (1985): Diagnosis of Alzheimer's disease. *Arch Neurol* 42: 1097–1105.

Kishikawa M, Iseki M, Nishimura M, Sekine I, Fujii H (1990): A histopathological study on senile changes in the human olfactory bulb. *Acta Pathol Jpn* 40:255–260.

Kobayashi K, Miyazu K, Fukutani Y, Nakamura I, Yamaguchi N (1991): Morphometric study on the CH4 of the nucleus basalis of Meynert in Alzheimer's disease. *Molec Chem Neuropathol* 15:193–206.

Lamy C, Duyckaerts C, Delaere P, Haus JJ (1989): Comparison of seven staining methods for senile plaques and neurofibrillary tangles in a prospective series of 15 elderly patients. *Neuropathol Appl Neurobiol* 15:563–578.

Lassmann H, Weiler R, Fischer P, Bancher C, Jellinger K, et al. (1992): Synaptic pathology in Alzheimer's disease. Immunological data for markers of synaptic and large dense-core vesicles. *Neuroscience* 46:1–8.

Mann DMA (1991): The topographic distribution of brain atrophy in Alzheimer's disease. *Acta Neuropathol* 83:81–86.

Mann DMA, Brown AMT, Prinja D, Jones D, Daviecs CA (1990): A morpholigical analysis of senile plaques in the brains of nondemented persons of different ages. *Neuropathol Appl Neurobiol* 16:17–25.

Mann DMA, Esiri MM (1989): The pattern of plaques and tangles in the brains of patients under 50 years of age with Down's syndrome. *J Neurol Sci* 89:169–179.

Mann DMA, Jones D, South PW, Snowden JS, Neary D (1992): Deposition of amyloid β protein in non-Alzheimer dementias: evidence for a neuronal origin of parenchymal deposits of β protein in neurodegenerative diseases. *Acta Neuropathol* 83:415–419.

Masliah E, Terry RD, Alford M, DeTeresa R, Hansen LA (1991a): Cortical and subcortical patterns of synaptophysin-like immunoreactivity in Alzheimer's disease. *Am J pathol* 138:235–246.

Masliah E, Hansen L, Albright T, Mallory M, Terry RD (1991b): Immunoelectron microscopic study of synaptic pathology in Alzheimer's disease. *Acta Neuropathol* 81:428–433.

McGeer PL, McGeer EG, Suzuki J, Doman CE, Nagai T (1984): Aging, Alzheimer's disease, and the cholinergic system of the basal forebrain. *Neurology* 34:743–745.

McKee AC, Kosik KS, Kowall NW (1991): Neuritic pathology and dementia in Alzheimer's disease. *Ann Neurol* 30:156–165.

Miyoshi K, Sato M (1991): Neuropathological and morphometric aspects of dementia, Alzheimer type. In: *Biological Psychiatry*, vol. 2, Racagni C, et al., eds. Amsterdam: Elsevier.

Morris JC, McKeel DW, Storandt M, Rubin EG, Price JL, Grant EA, Ball MJ, Berg L (1991): Very mild Alzheimer's disease. *Neurology* 41:469–478.

Mountjoy CQ (1988): Number of plaques and tangles, loss of neurons: their relation with deficient neurotransmitter synthesis and the degree of dementia. *Interdiscipl Topics Gerontol* 25:74–89.

Murphy GM, Ellis WG (1991): The amygdala in Down's syndrome and familial Alzheimer's disease. *Biol Psychiatr* 30:92–106.

Nakamura S, Akiguchi I, Kameyama M, Mizuno N (1985): Age-related changes of pyramidal cell basal dendrites in layers III and V of human motor cortex. *Acta Neuropathol* 65:281–284.

Ogomori K, Kitamoto T, Tateishi J, Sato Y, Suetsugu M, Abe M (1989): β-protein amyloid is widely distributed in the central nervous system of patients with Alzheimer's disease. *Am J Pathol* 134:243–251.

Pappola MA, Omar RA, Vinters HV (1991): Image analysis microspectoscopy shows that neurons participate in the genesis of a subset of early primitive (diffuse) senile plaques. *Am J Pathol* 139:599–607.

Paulus W, Bancher C, Jellinger K (1992): Interrater reliability in the neuropathologic diagnosis of Alzheimer's disease. *Neurology* 42:329–332.

Perry EK, Smith CJ, Perry RH, et al. (1989): Nicotinic (^{3}H-nicotine) receptor binding in human brain. *Neurosci Res Commun* 5:117–124.

Podlisny MB, Tolan DR, Selkoe DJ (1991): Homology of the amyloid β-protein precursor in monkey and human supports a primate model for beta amyloidosis in Alzheimer's disease. *Am J Pathol* 138:1423–1435.

Powers JM, Skeen JT (1988): Ultrastructural heterogeneity in cerebral amyloid of Alzheimer's disease. *Acta Neuropathol* 77:168–174.

Price LJ, Davis PB, Morris JC, White DL (1991): The distribution of tangles, plaques and related immunohistological markers in healthy aging and Alzheimer's disease. *Neurobiol Aging* 12:295–312.

Probst A, Langui D, Ipsen S, Robakis N, Ulrich J (1991): Deposition of β/A4 protein along neuronal plasma membranes in diffuse senile plaques. *Acta Neuropathol* 83:21–29.

Quon D, Wang Y, Catalano R, Scardina JM, Murakami K, Cordell B (1991): Formation of beta-amyloid protein deposits in brains of transgenic mice. *Nature* 352:239–241.

Ransmayer G, Cervera P, Hirsch E, Ruberg M, et al. (1989): Choline acetyltransferase-like immunoreactivity in the hippocampal formation of control subjects and patients with Alzheimer's disease. *Neuroscience* 32:701–714.

Rogers J, Styren SC (1987): Neuroanatomy of aging and dementia. *Rev Biol Res Aging* 3:223–253.

Sandhu FA, Salim M, Zain SB (1991): Expression of the human beta-amyloid protein of Alzheimer's disease specifically in the brains of transgenic mice. *J Biol Chem* 266:21331–21334.

Scheff SW, DeKosky ST, Price DA (1990): Quantitative assessment of cortical synaptic density in Alzheimer's disease. *Neurobiol Aging* 11:29–37.

Scheibel AB (1979): Aging in human motor control system. *Aging* 10:297–310.

Schubert W, Prior R, Wedemann A, Dircksen H, Multhaupt G, Masters CL, Beyreuther K (1991): Localization of Alzheimer beta A4 amyloid precursor protein at central and peripheral synaptic sites. *Brain Res* 563:184–194.

Scott SA, DeKosky ST, Sparks LD, Knox CA, Scheff SW (1992): Amygdala cell loss and atrophy in Alzheimer's disease. *Ann Neurol* 32:555–563.

Selkoe DJ (1991a): The molecular pathology of Alzheimer's disease. *Neuron* 6:487–498.

Selkoe DJ (1991b): Alzheimer's disease — in the beginning. *Nature* 354:432–433.

Shivers BD, Hilbich C, Multhaup G, Salbaum M, Beyreuther K, Seeburg PH (1988): Alzheimer's disease amyloidogenic glycoprotein: expression pattern in rat brain suggest a role in cell contact. *EMBO J* 7:1365–1370.

Struble RG, Clark HB (1992): Olfactory bilb lesions in Alzheimer's disease. *Neurobiol Aging* 13:469–473.

Swaab DF, Grundke-Iqbal I, Iqbal K, Kremer HPH, Ravid R, Van de Nes JAP (1992): Tau and ubiquitin in the human hypothaalmus in aging and Alzheimer's disease. *Brain Res* 590:239–249.

Tagliavani F, Ghiso J, Timmers WF, Giaccone G, Bugiani O, Frangione B (1990): Coexistence of Alzheimer's amyloid precursor protein and amyloid protein in cerebral vessel walls. *Lab Invest* 62:761–767.

Tagliavani F, Giaccone G, Frangione B, Bugiani O (1988): Preamyloid deposits in the central cortex of patients with Alzheimer's disease and nondemented individuals. *Neurosci Lett* 93:191–196.

Talamo BR, Rudel RA, Kosik K, Lee VMY, Neff S, Edelman L, Kauer JS (1989): Pathology in olfactory neurons of Alzheimer's disease. *Nature* 37:736–739.

Terry RD, DeTeresa R, Hansen RA (1987a): Neocortical cell counts in normal human adult aging. *Ann Neurol* 21:530–539.

Terry RD, Hansen LA, DeTeresa R, et al. (1987b): Senile dementia of the Alzheimer type without neocortical neurofibrillary tangles. *J Neuropathol Exp Neurol* 46:262–268.

Terry RD, Masliah E, Salmon DP, Butters N, DeTeresa R, Hill R, Hansen LA, Katzman R (1991): Physical basis of cognitive alteration in Alzheimer's disease: synapse loss is the major correlate of cognitive impairment. *Ann Neurol* 30:572–580.

Tierney MC, Fischer H, Lewis AJ, et al. (1988): The NINCDS-ADRDA Work Group criteria for the clinical diagnosis of probable Alzheimer's disease: clinicopathologic study of 57 cases. *Neurology* 38:356–364.

Tomlinson BE, Henderson G (1976): Some quantitative cerebral findings in normal and demented old people. In *Neurobiology of Aging*, Terry RD, Gershon S, eds. New York: Raven Press.

Tomlinson BE (1989): The neuropathology of Alzheimer's disease. Issues in need of resolution. *Neuropathol Appl Neurobiol* 15:491–512.

Trojanowski JQ, Newman PD, Hill WD, Lee VMY (1991): Human olfactory epithelium in normal aging, Alzheimer's disease, and other neuro-degenerative disorders. *J Comp Neurol* 310:365–376.

Uchihara T, Kondo H, Kosaka K, Tshukagoshi H (1992): Selective loss of nigral neurons in Alzheimer's disease: a morphometric study. *Acta Neuropathol* 83.

Ulrich J (1985): Alzheimer changes in nondemented patients younger than sixty-five: possible early stages of Alzheimer's disease and senile dementia of Alzheimer type. *Ann Neurol* 17:273–277.

Ulrich J (1988): Histology and Histopathology of the Aging Brain. *Interdisc Topics Gerontol*, Vol. 25; Basel: Karger.

Unger JW, Lange W (1992): NADPH-diaphorase positive cell populations in the human amygdala and temporal cortex: neuroanatomy, peptidergic characteristics and aspects of aging and Alzheimer's disease. *Acta Neuropathol* 83:636–646.

Unger JW, Lapham LW, McNeill TH, Eskin TA, Hamill RW (1988): The amygdala in Alzheimer's disease: neuropathology and Alz-50 immunoreactivity. *Neurobiol Aging* 12:389–399.

Vallet PG, Guntern R, Hof PR, Golaz J, Delacourte A, Robakis NK, Bouras C (1992): A comparative study of histological and immunohistochemical methods for neurofibrillary tangles and senile plaque in Alzheimer's disease. *Acta Neuropathol* 83:170–178.

Vogels OJM, Broere AJ, Ter Laak HJ, Ten Doneklar HJ, Nieuwenhuis R, Schulte BPM (1990): Cell loss and shrinkage in the nucleus basalis Meynert complex in Alzheimer's disease. *Neurobiol Aging* 11:3–13.

Weis S, Jellinger K, Wenger E (1991): Morphometry of the corpus callosum in normal aging and Alzheimer's disease. *J Neural Transm* 33(Suppl):35–38.

Wirak DO, Bayney B, Ramabhadran TV, Fracasso RP, et al. (1991): Deposits of amyloid-β protein in the central nervous system of transgenic mice. *Science* 253:323–325.

Wisniewski HM, Wen GY (1985): Substructures of paired helical filaments from Alzheimer's disease fibrillary tangles. *Acta Neuropathol* 66:173–176.

Wisniewski HM, Wen GY (1988): Lipopigment in the aging brain. *Am J Med Genet.* 5(Suppl):183–191.

Wisniewski HM, Wen GY, Kim LS (1989): Comparison of four staining methods on the detection of senile plaques. *Acta Neuropathol* 78:22–27.

Wisniewski HM, Wegiel J, Morys J, Bancher C, Soltysiak Z, Kim KS (1990): Aged dogs: an animal model to study beta-protein amyloidogenesis. In: *Alzheimer's Disease: Epidemiology, Neuropathology, Neurochemistry, and Clinics*, Maurer K, et al., eds. Vienna: Springer-Verlag.

Yamaguchi H, Nakazato Y, Shoji M, Takatama M, Hirai S (1991): Ultrastructure of diffuse senile plaques in senile dementia of the Alzheimer type. Comparison with primitive plaques. *Acta Neuropathol* 82:13–20.

Yates CM, Simpson J, Gordon A (1990): 5-hydroxytryptamine and dopamine in pre-senile and senile Alzheimer-type dementia. *Biochem Soc Trans* 18:422–423.

Xuereb JH, Perry RH, Candy CM, Perry EK, Marshall E, Bonham JR (1991):

Nerve cell loss in the thalamus in Alzheimer's disease and Parkinson's disease. *Brain* 114:1363–1379.

Zubenko GS, Moossy J, Kopp U (1990): Neurochemical correlates of major depression in primary dementia. *Arch Neurol* 47:209–214.

Zubenko GS, Moossy J, Martinez AJ, et al. (1989): A brain regional analysis of morphologic and cholinergic abnormalities in Alzheimer's disease. *Arch Neurol* 46:634–638.

Zweig Rm, Rose CA, Hedreen JC, Steele C, Pride DL (1988): The neuropathology of aminergic nuclei in Alzheimer's disease. *Ann Neurol* 24:233–242.

Chapter 11

Natural History of Alzheimer's Disease

Heiko Braak and Eva Braak

The hallmark of Alzheimer's disease is a progressive accumulation of insoluble fibrous pathological material in both extracellular and intraneuronal locations. The extracellular deposits are formed of the A4 amyloid protein (Joachim and Selkoe, 1989), while the intraneuronal material is largely made up of abnormally phosphorylated tau protein representing the neurofibrillary changes of the Alzheimer type (i.e., neuritic plaques, neurofibrillary tangles, neuropil threads) (Gambetti et al., 1990; Iqbal and Grundke-Iqbal, 1991; Braak and Braak, 1991b; Perry et al., 1991).

The distribution pattern of amyloid deposits does not correlate to that of neurofibrillary changes. Frequently, amyloid deposits are found in the absence of neurofibrillary changes but the inverse pattern (i.e., neurofibrillary changes in the absence of amyloid deposits) occurs as well. Amyloid deposits, therefore, cannot be considered a prerequisite for the development of neurofibrillary changes (Braak and Braak, 1991b). Amyloid deposits are first seen in low densities in basal portions of the isocortex. Next affected are the presubiculum and entorhinal cortex. Only the more advanced cases show amyloid deposits in the hippocampal formation as well. The distribution pattern and packing density of amyloid deposits is subject to considerable interindividual variation. Analysis of amyloid distribution, therefore, is of only limited significance for differentiation of stages of Alzheimer-related pathological changes (Braak and Braak, 1991b).

A reliable neuropathological staging is likewise not possible when based on the distribution pattern of neuritic plaques. Inconsistent presence, varying density of neuritic plaques even within defined areas, and their considerable interindividual variation hamper any such attempt. In

ADVANCES IN RESEARCH ON NEURODEGENERATION, I
D.B. Calne et al.

contrast, neurofibrillary tangles and neuropil threads exhibit a selective affection of specific architectonic units and display a characteristic sequence of changes during the course of Alzheimer's disease (Hyman et al., 1990; Arnold et al., 1991; Braak and Braak, 1991b; Price et al., 1991).

The most severe lesions observable in Alzheimer's disease generally are seen in limbic areas of the cerebral cortex such as the hippocampal formation and the entorhinal region. The entorhinal region exhibits a particularly complex lamination pattern and a large number of architectonic units. Up to 11 layers can be identified. However, neither its layers nor its cellular constituents are comparable to those of the isocortex (Braak, 1980; Braak and Braak, 1985). Large nerve cells residing within the cellular islands of the superficial entorhinal layer give rise to major portions of the perforant path, an important fiber tract of the limbic system that connects the entorhinal region with the hippocampal formation (Fig. 1). The transentorhinal region represents a transition zone between the entorhinal allocortex and the temporal isocortex. Its distinguishing feature is the gradual descent of the superficial entorhinal layer following an oblique course through the outer parts of the cortex (Fig. 2) (Braak, 1980; Braak and Braak, 1985).

In lower mammals it is mainly olfactory information that is processed in the entorhinal cortex. With phylogenetic advance, however, the influence of the limbic thalamus and other relay stations of the limbic system is considerably enlarged and eventually afferents from the isocortex predominate. In higher primates and in man, the transentorhinal region serves as the gate of entrance for isocortical information. Isocortical association areas project via large numbers of cortical relay stations on the transentorhinal and entorhinal regions, thereby providing them with abundant somatomotor, somatosensory, acoustic, and visual information. Via the presubiculum there is an additional strong input from the anterior nuclear complex of the thalamus and from the reuniens nucleus, both important centers of the limbic system. The perforant path eventually transports both limbic and isocortical data to the hippocampus (Fig. 1). This continuous information transfer from limbic circuits and isocortex to the hippocampal formation is of utmost significance for maintaining mnemonic functions.

Observations of neurofibrillary changes in the brains of large numbers of mildly and severely affected patients can be synthesized into a probable sequence of events. This sequence is conceptualized as six

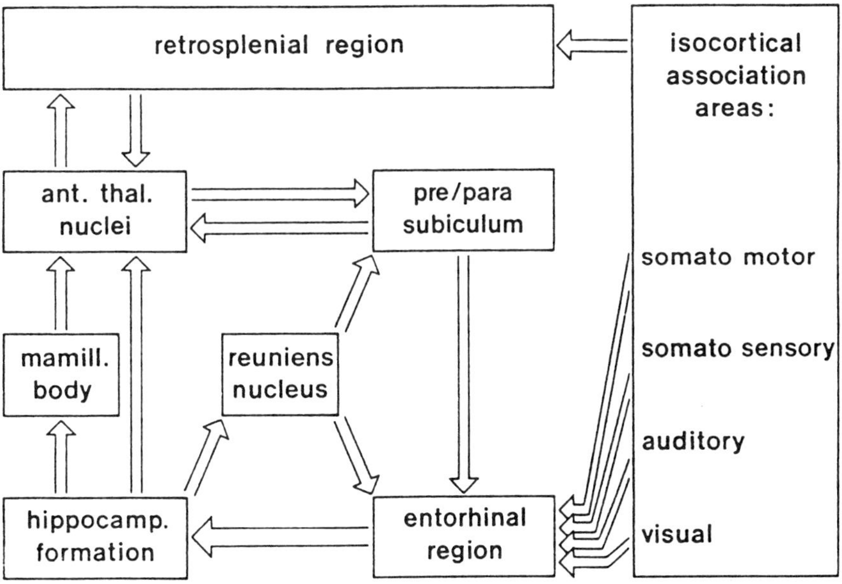

Figure 1 Via numerous cortical relay stations isocortical association areas provide the entorhinal region with abundant somatomotor, somatosensory, acoustic, and visual informations. Via the presubiculum there is additional input from the anterior nuclei of the thalamus and from the reuniens nucleus, both important centers of the limbic system.

stages of a pathological process. Each set of changes may account for different clinical aspects of Alzheimer's disease.

The most mildly affected cases consistently show the involvement of only the superficial entorhinal cellular layer in the transentorhinal region. The projection cells in this layer and in this area are the first neurons in the entire brain that develop neurofibrillary tangles and neuropil threads. Stage I, accordingly, is defined by the presence of no more than a few tangles and threads in the transentorhinal region, whereas stage II already exhibits numerous tangles and threads in this location. At stage II, some neurofibrillary tangles also may occur in the proper entorhinal region and in the hippocampal CA1 sector.

Stages III and IV exhibit already subcortical changes in a few nuclei of the limbic system. The reuniens nucleus and, in particular, the anterodorsal nucleus of the thalamus show the presence of numerous neurofibrillary tangles and neuropil threads (Braak and Braak, 1991a). The characteristic feature of stage III, however, is the severe involvement of the superficial entorhinal cellular layer, in both the transentorhinal and

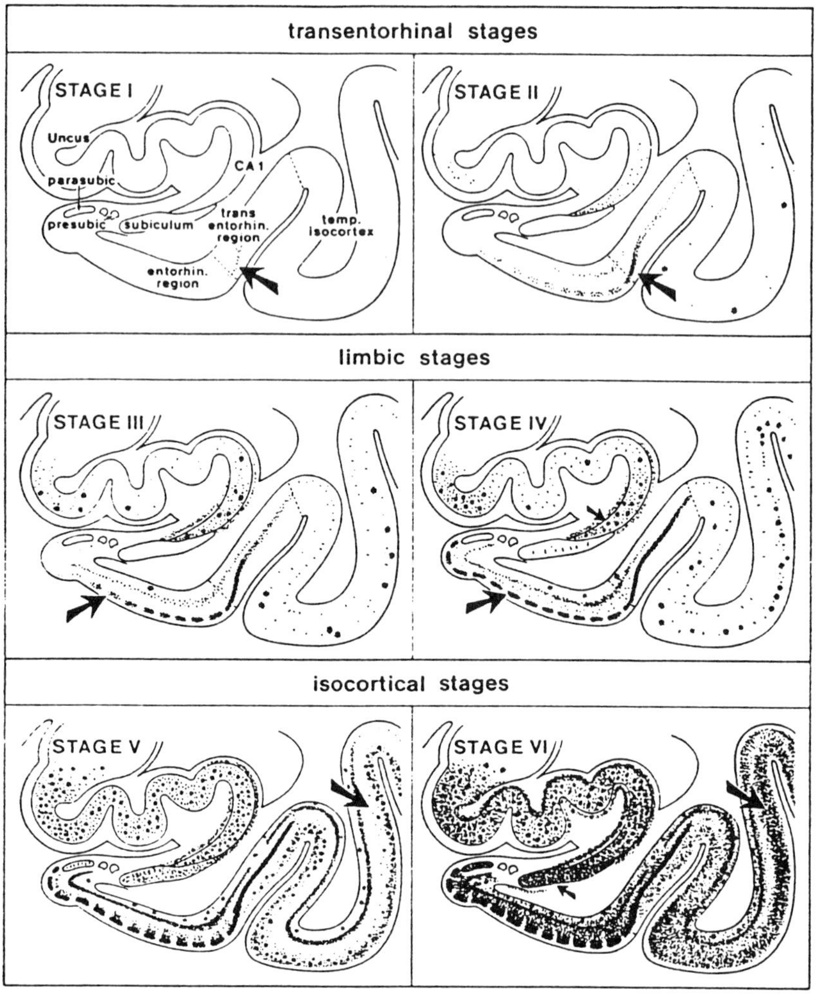

Figure 2 Summary diagram of neurofibrillary changes seen in anteromedial portions of the temporal lobe and development of changes from stage I to stage VI. The arrows point to leading characteristics. Areas are outlined in stage I (CA1: sector 1 of the Ammon's horn). Reprinted with permission of Springer-Verlag Heidelberg from Braak H, Braak E (1991b): Neuropathological staging of Alzheimer-related changes. *Acta Neuropathol* 82:239–259.

entorhinal regions; the hippocampal formation and the isocortex still remain unaffected or are only mildly involved (Braak and Braak, 1990a,b). At stage IV, one of the deep entorhinal cellular layers shows a marked involvement and isocortical association areas adjoining the allocortex develop modestly dense neurofibrillary changes (Fig. 2).

At stage V, the hippocampal CA1 sector is filled with neurofibrillary tangles and dense networks of neuropil threads appear above and below the pyramidal cell layer. The leading feature of stages V and VI, however, is the severe effect on the isocortical association areas, leaving only primary motor and primary sensory areas uninvolved or mildly affected. Primary sensory areas initially develop neuritic plaques in supragranular layers and, as a special feature, a dense network of neuropil threads in layer V (Braak et al., 1989). At the final stage, VI, even primary sensory areas show a large number of neurofibrillary tangles in layers II and III, and a sharply demarcated layer V tightly filled with neuropil threads.

In summary, stages I and II are characterized by an accentuated involvement of the transentorhinal region and they are therefore referred to as the "transentorhinal stages." The clinical protocols of cases classified as belonging to stages I or II do not include overt impairment of intellectual capacities. They are therefore considered to represent the clinically silent period or preclinical phase of Alzheimer's disease.

The key feature of stages III and IV is the severe involvement of the superficial entorhinal cellular layer in both the transentorhinal and entorhinal region. This is supplemented by involvement of a few limbic subcortical nuclei. The stages are therefore summarized as "limbic stages." The clinical protocols of cases in stages III or IV note an impairment of cognition and slight personality changes. The limbic stages III and IV probably represent clinically incipient Alzheimer's disease (Braak and Braak, 1990a,b, 1991b).

The common hallmark of stages V and VI is the devastating involvement of the isocortex, in particular the destruction of isocortical association areas. Stages V and VI are therefore denoted as "isocortical stages." The clinical protocols of all cases diagnosed as isocortical stages V or VI note the presence of dementia. These stages also meet the conventional criteria for neuropathological confirmation of the clinical diagnosis of Alzheimer's disease (Khachaturian, 1985; Jellinger et al., 1991) (i.e., they correlate with fully developed dementia of the Alzheimer type).

The key characteristic of the pathological process underlying initial Alzheimer's disease is a bilateral destruction of only a few specific areas, layers, and cell types. From stage I to IV, only a small percentage of the total nerve cell number of the brain is affected. All of the affected structures are tightly interconnected components of the limbic system. Because of the key position of the entorhinal region within the limbic system, a bilateral involvement of only its superficial cellular layer already causes severe functional disturbances and particularly hampers the transmission of information from the isocortex to the hippocampal formation (Hyman et al., 1984, 1990; Braak and Braak, 1985; Van Hoesen et al., 1991). Additional destruction of nuclei of the limbic thalamus leads to disturbances of limbic circuits at multiple sites. These lesions most likely contribute to the early cognitive decline seen in individuals suffering from Alzheimer's disease. Substantial loss of nerve cells is noted only in the end stages of Alzheimer's disease.

Acknowledgments. This study was supported by the Deutsche Forschungsgemeinschaft and Degussa, Hanau. The authors would also like to thank Mrs. Szasz for the drawings and Mr. Müller for critical reading of the manuscript.

References

Arnold SE, Hyman BT, Flory J, Damasio AR, van Hoesen GW (1991): The topographical and neuroanatomical distribution of neurofibrillary tangles and neuritic plaques in the cerebral cortex of patients with Alzheimer's disease. *Cerebral Cortex* 1:103–116.

Braak H (1980): Architectonics of the human telencephalic cortex. In: *Studies of Brain Function, Vol. 4*, Braitenberg V, Barlow HB, Bizzi E, Florey E, Grüsser OJ, van der Loos E, eds. Berlin: Springer.

Braak H, Braak E (1985): On areas of transition between entorhinal allocortex and temporal isocortex in the human brain. Normal morphology and lamina-specific pathology in Alzheimer's disease. *Acta Neuropathol* 168:325–332.

Braak H, Braak E (1990a): Cognitive impairment in Parkinson's disease: amyloid plaques, neurofibrillary tangles and neuropil threads in the cerebral cortex. *J Neural Transm(P-D Sect)* 2:45–57.

Braak H, Braak E (1990b): Neurofibrillary changes confined to the entorhinal region and an abundance of cortical amyloid in cases of presenile and senile dementia. *Acta Neuropathol* 80:479–486.

Braak H, Braak E (1991a): Alzheimer's disease affects limbic nuclei of the thalamus. *Acta Neuropathol* 81:261–268.

Braak H, Braak E (1991b): Neuropathological stageing of Alzheimer-related changes. *Acta Neuropathol* 82:239–259.

Braak H, Braak E, Kalus P (1989): Alzheimer's disease: areal and laminar pathology in the occipital isocortex. *Acta Neuropathol* 177:494–506.

Gambetti P, Tabaton M, Cammarata S, Morandi A, Schaetzle B, Wicker N, Perry G, Autilio-Gambetti L (1990): Neurofibrillary tangles, neuropil threads and neuritic plaques in Alzheimer's disease: are these lesions interrelated. In: *Molecular Biology and Genetics of Alzheimer's Disease*, Miyatake T, Selkoe DJ, Ihara Y, eds. Amsterdam: Elsevier, pp. 57–66.

Hyman BT, van Hoesen GW, Damasio AR, Barnes CL (1984): Alzheimer's disease: cell-specific pathology isolates the hippocampal formation. *Science* 225:1168–1170.

Hyman BT, van Hoesen GW, Damasio AR (1990): Memory-related neural systems in Alzheimer's disease: an anatomic study. *Neurology* 40:1721–1730.

Iqbal K, Grundke-Iqbal I (1991): Alzheimer's disease: from cytoskeletal protein pathology to neuronal degeneration. In: *Alzheimer's Disease: Basic Mechanisms, Diagnosis and Therapeutic Strategies*, Iqbal K, McLachlan DRC, Winblad B, Wisniewski HM, eds. Chichester: Wiley, pp. 173–180.

Jellinger K, Lassmann H, Fischer P, Danielczyk W (1991): Validation of diagnostic criteria for Alzheimer's disease. In: *Alzheimer's Disease: Basic Mechanisms, Diagnosis and Therapeutic Strategies*, Iqbal K, McLachlan DRC, Winblad B, Wisniewski HM, eds. Chichester: Wiley, pp. 75–89.

Joachim CL, Selkoe DJ (1989): Minireview: Amyloid protein in Alzheimer's disease. *J Gerontol (Biol Sci)* 44:77–82.

Khachaturian ZS (1985): Diagnosis of Alzheimer's disease. *Arch Neurol* 142: 1097–1105.

Perry G, Kawai M, Tabaton M, Onorato M, Mulvihill P, Richey P, Morandi A, Connolly JA, Gambetti P (1991): Neuropil threads of Alzheimer's disease show a marked alteration of the normal cytoskeleton. *J Neurosci* 11:1748–1755.

Price JL, Davies PB, Morris JC, White DL (1991): The distribution of tangles, plaques and related immunohistochemical markers in healthy aging and Alzheimer's disease. *Neurobiol Aging* 12:295–312.

Van Hoesen GW, Hyman BT, Damasio AR (1991): Entorhinal cortex pathology in Alzheimer's disease. *Hippocampus* 1:1–8.

Part IV

Amyotrophic Lateral Sclerosis

Chapter 12

Amyotrophic Lateral Sclerosis: Clinical Definition and Relation to Other Neurodegenerative Disorders

C. W. Olanow

Amyotrophic lateral sclerosis (ALS) or motor neuron disease can be defined as a progessive, degenerative disorder of motor neurons in the spinal cord, brain stem, and motor cortex characterized by varying degrees and patterns of weakness, atrophy, fasciculations, and corticospinal tract signs. Like Parkinson's disease (PD) and Alzheimer's disease (AD), it is a disease of the later years of life in which selected subsets of neurons appear to be specifically vulnerable to a degenerative process. The appreciation that ALS is related to degeneration of anterior horn cells is attributed to the work of Charcot and his students in the mid-19th century. Interestingly, unlike Alzheimer's and Parkinson's diseases, no eponym is associated with its title in the English-speaking world. There is considerable variability in the clinical presentation of ALS and the full extent of the disease spectrum remains to be fully defined (World Federation of Neurology, 1988). Because of a lack of a clear etiology or pathogenesis, ALS is generally classified according to clinical presentation and whether it is sporadic or familial (Williams and Windebank, 1991).

A number of different clinical syndromes fall under the umbrella of the term ALS based on the site of involvement and the presence or absence of upper or lower motor neuron dysfunction (Table 1). Ultimately, most cases progress to develop a clinical picture in which there is combined upper and lower motor neuron involvement. In the United States, the term ALS is used to encompass all of these syndromes but in Great Britain the more generic term motor neuron disease is employed and ALS is reserved only for those syndromes in which both upper and lower mo-

ADVANCES IN RESEARCH ON NEURODEGENERATION, I
D.B. Calne et al.

Table 1. Clinical Variants of the Amyotrophic Lateral Sclerosis (ALS) Syndrome

Area of Involvement		Syndrome
Bulbar	Upper motor neuron	Pseudobulbar palsy
	Lower motor neuron	Bulbar palsy
Spinal	Upper motor neuron	Primary lateral sclerosis
	Lower motor neuron	Progressive Spinomuscular atrophy
Combination of above		Amyotrophic lateral sclerosis

tor neuron features are present. There are some who argue that patients with features confined to lower motor neurons should be considered as having a separate disease entity (Norris et al., 1978). But in these patients pathology is rarely confined to lower motor neurons and it is likely that this syndrome is a part of the ALS spectrum. On the other hand, clinical involvement confined to upper motor neurons (primary lateral sclerosis) is less clearly a part of ALS and might represent a different disease entity (Younger et al., 1988). While there are a range of clinical syndromes comprising ALS, clinicians rarely disagree on diagnosis (Rowland, 1980; Buckley et al., 1985) and errors typically represent a failure to diagnose in the elderly (Brownell et al., 1970; Bradley, 1987).

ALS has a global distribution and is estimated to have an incidence of approximately 2/100,000/year, with a male proponderance (Murros and Fogelholm, 1983; Yoshida et al., 1986; Kondo and Tsubaki, 1977). Lower incidence rates probably reflect incomplete ascertainment (Kahana and Zilber, 1984; Durrleman and Alperovitch, 1989). ALS can present over a wide variety of age groups but onset is usually in the sixth decade and there is an increasing age-specific risk of acquisition (Kondo, 1979). There is also an age-specific increase in mortality rate that is higher in recent decades, probably reflecting more accurate diagnosis and more complete ascertainment (Lilienfeld et al., 1989). Interestingly, identical observations have been made with PD. Numerous epidemiologic studies have sought risk factors for ALS. Trauma is found almost consistently in both prospective (Kurtzke and Beebe, 1980) and retrospective (Deapen and Henderson, 1986) studies. Sporadic, familial and Western Pacific forms of the disease have been described. Most cases are sporadic.

Approximately 5% to 10% are familial and are generally transmitted according to an autosomal dominant pattern of inheritance (Kurland and Mulder, 1985; Mulder et al., 1986; Emery and Holloway, 1982). Patients with familial forms of ALS tend to be younger, to have onset in the lower extremities, to be more likely to have other systems involved, and to progress more rapidly (Emergy and Holloway, 1982; Mulder et al., 1986; Hirano et al., 1967). Variability of clinical expression can be seen in and between individual families in both clinical (Chio et al., 1987) and pathological (Tanaka et al., 1984; Engel et al., 1959; Hirano et al., 1967) features although most have patterns that resemble those found in classic sporadic ALS (Bonuelle, 1975). The ALS syndrome found in the Western Pacific is clinically identical but is associated with widespread neurofibrillary tangle (NFT) formation and often with parkinsonism and dementia (Hirano et al., 1961a,b). The incidence of the ALS component of this syndrome appears to be declining, although the parkinsonian and dementing components persist (Kurland and Mulder, 1988).

Weakness (+/− atrophy and fasciculations) is most often the presenting feature of ALS (Gubbay et al., 1985). The limbs tend to be involved before the bulb (Vejjajiva et al., 1967) and the upper extremities before the lower extremities (Li et al., 1988). Initial presentation is usually asymmetric (Swash and Schwartz, 1982) and at times can be focal (Munsat et al., 1988). Cramps, fatigue, and nonspecific sensory complaints can be features. It is however diffuse weakness, atrophy, and fasciculations that are the signature of the disease and the clinical features at any one time in the illness reflect the site of involvement and the presence or absence of upper and lower motor neuron features. Intellect, sphincter control, and sensory functions are traditionally preserved (Gubbay et al., 1985). However, sensory complaints are frequently reported (Lawyer and Netsky, 1953; Dyck et al., 1975; Tandan and Bradley, 1985) and are detected with quantitative techniques in as many as 15% of patients (Mulder et al., 1983). Eye muscles are generally spared, although in advanced respirator-dependent cases, involvement of external ocular muscles can occur (Iwata, 1987). Autonomic dysfunction also can be detected in many patients if carefully sought (Chida et al., 1989).

ALS is almost universally fatal but progression is somewhat variable (Gubbay et al., 1985) and in rare instances cessation and even reversal of progression have been recorded (Mulder and Howard, 1976). The disease tends to spread first at the level of initial involvement (i.e., from one arm to the other or from one leg to the other). Ascending involvement is more typical than descending progression. Patients with bulbar involvement tend to progress most rapidly and those with involvement confined to the lower extremities most slowly. Death typically ensues within 2 to 4 years from the time of diagnosis, generally as a result of aspiration or respiratory infection (Gubbay et al., 1985).

There is no specific laboratory markers for ALS and diagnosis remains primarily clinical. Symptomatic forms of motor neuron dysfunction must be excluded but are uncommon. ALS-like syndromes have been reported in association with plasma cell dyscrasias, hemoglobin deficiency, hypoglycemia, heavy metals, thyroid disease, cancer, and in patients with antecedent polio (Williams and Windebank, 1991). Electrophysiological studies are helpful and findings of diffuse denervation in territories innervated by multiple peripheral nerves or roots with relatively preserved motor and sensory nerve conductions support a diagnosis of ALS (Bradley, 1987). Swash and Schwartz (1984) have defined criteria for staging ALS based on clinical and neurophysiologic criteria.

It is also likely that there is a preclinical syndrome of ALS. As with PD, aging is associated with a gradual loss of anterior horn cells (Howard et al., 1988) and as many as 70% to 80% must be lost before clinical weakness can be appreciated or detected. Denervation changes can be detected in clinically unaffected muscle groups in ALS patients. Lou Gehrig, perhaps the most famous victim of this disease, was diagnosed as having ALS in 1939, but a review of his previous stable and predictable performance demonstrates a dramatic deterioration in batting average, RBIs, and home runs in 1938 before subjective complaints or diagnosis, reflecting the probability that he had clinically unappreciated ALS at that time (Kasarkis and Winslow, 1989).

It is also noteworthy that overlap syndromes occur with the other neurodegenerative disorders. Patients with ALS are reported to have an increased incidence of dementia and parkinsonism (Hudson, 1981; Calne and Eisen, 1989) and motor neuron dysfunction can be seen in association with dementing illnesses and parkinsonism (Gajdusek and Salazar, 1982; Shy and Drager, 1960; Ishino et al., 1974; Delisle et al., 1987). Others have maintained that a relationship between Parkinson's disease and Alzheimer's disease is coincidental (Kondo, 1984; Leone et al., 1987). However, the findings of diffuse, non-Alzheimer neurodegeneration in patients with ALS syndromes suggest that it is either a part of the spectrum of ALS (Hudson, 1981) or a distinct entity (Mitsuyama, 1984). Patients with ALS have reduced 2-deoxy-D-glucose utilization in the basal ganglia (P. McGeer, *personal communication*) and signal hypointensity in the putamen on high field strength magnetic resonance imaging (MRI) studies, as is found in patients with multisystem atrophy (Olanow and Drayer, 1987). Biopsies of clinically unaffected muscles from patients with Parkinson's disease demonstrate denervation (M. Dalakis, *personal communication*). These observations suggest that motor neuron involvement can be a component of multisystem degeneration and that in at least some cases there may be a common underlying etiology or pathogenesis.

Pathologically, ALS is characterized by a dropout of large motor neurons and varying degrees of gliosis and degeneration of residual neurons as well as ghost cells, and neuronophagia (Chou, 1979; Hughes, 1982; Hirano and Iwata, 1979). Lower brain stem nuclei tend to be more involved than those in the upper brain stem and cervical anterior horn cells more than those in the lumbar region (Tsukagoshi et al., 1988). Sacral neurons tend to be spared (Katagiri et al., 1988) except in late respirator-dependent cases (Iwata, 1987). It is unknown why cells then can participate in the degenerative process are less vulnerable than others. There is usually a significant loss of Betz cells and to a lesser degree of pyramidal cells from precentral and postcentral motor cortex (Hammer et al., 1979). Additional findings include a loss of large myelinated fibers in peripheral nerves and roots (Sobue et al., 1981) and demyelination of the spinal cord in the region of the corticospinal tracts, but typically with sparing of the posterior columns (Lawyer and Netsky, 1953). Although the sensory examination tends to be normal, sensory fibers and ganglia can be involved (Bradley, et al., 1983), as can nonmotor cortical and subcortical systems, including multisystem involvement in at least some patients.

A number of cytoskeletal and ultrastructural features are also observed (Hirano and Iwata, 1979), although none are pathognomonic. Spheroids, comprised of neurofilaments, are observed more frequently in proximal axons of anterior horn cells of ALS patients than in controls (Delisle and Carpenter, 1984). Two- to three-micron eosinophilic inclusion bodies (Bunina bodies) may be found in the cytoplasm of anterior horn cells in patients with ALS. It is noteworthy that ubiquitin-staining deposits, which are found in neurofibrillary tangles in AD and Lewy bodies in PD (Perry et al., 1987; Kuzuhara et al., 1988), also are found in anterior horn cells (Leigh et al., 1988; Lowe et al., 1989), brain stem, and motor cortex of patients with ALS but not controls.

One is struck by the many similarities between ALS and the other neurodegenerative disorders such as PD and AD (Table 2). Each is a disorder of specific sets of aging neurons that appear to be selectively vulnerable to a degenerative process of unknown etiology and pathogenesis. Familial cases comprise 5% to 10% of cases and these patients tend to be younger, clinically atypical, and progress at faster rates. This suggests that genetic factors play a role in at least subsets of populations with these diseases. In each, there is considerable neuronal reserve before the appearance of clinical dysfunction and in each degeneration occurs with either no or minimal detectable inflammatory or vascular change. In each, the pathology appears to be largely confined to the degenerating neurons that contain ubiquitin-staining cytoskeletal products in the form of inclusion bodies or tangles. Each has an age of onset in and around the 6th or 7th decade. Both ALS and Parkinson's disease are said to have a strong correlation between incidence and latitude, with an increasing frequency at latitudes remote from the equator. All invariably progress and ultimately lead to death. Death rate increases with increasing age and there is a striking similarity in age-specific death rate by age and decade for at least patients with ALS and PD (Lilienfeld et al., 1989). Each of these degenerative disorders can be seen in a "pure" form in which there is almost universal agreement as to diagnosis by clinicians and pathologists. In contrast, clinical and pathologic variants are described that challenge our definition of the clinical spectrum and in which the relation to the parent syndrome is uncertain.

Clinical and pathologic overlap among the neurodegenerative disorders further confuses the issue. Fragments of ALS, Parkinson's disease, Alzheimer's disease, and multisystem atrophy can be observed within a single individual (Delisle et al., 1987). Patients with Parkinson's disease develop clinical dementia in as many as one-third of cases and more have clinically undetected Alzheimer pathology (Hakim and Mathieson, 1979; Lieberman et al., 1979; Jellinger, 1987; Mayeux et al., 1986). Patients with Alzheimer's disease frequently have parkinsonian features (Molsa et al., 1984; Pearce, 1974; Leverenz and Sumi, 1986; Ditter and Mirra, 1987). Both dementia and parkinsonism can be seen in association with motor neuron dysfunction (Neary et al., 1990; Salazar et al., 1983) and patients with multisystem atrophy also can have motor neuron dysfunction (Shy and Drager, 1960; Ishino et al., 1974; Delisle et al., 1987). Pathologically, each is associated with ubiquitin-positive inclusions or tangles comprised of cytoskeletal elements and phosphorylated neurofilaments are abnormally increased in ALS (Sobue et al., 1990),

Table 2. Features of Neurodegenerative Disorders

Age-related decline in specific set of selectively vulnerable neurons
Inexorable progression
Etiology and pathogenesis unknown
Degeneration of majority of neurons occurs before clinical features develop
Males more affected than females (PD and ALS)
Age of onset tends to be in the sixth or seventh decade but may be wide variation
Correlation between incidence and latitude (PD/ALS)
Well defined clinical syndrome but extent of clinical syndrome is not established
Variable patterns of clinical presentation
Can have atypical presentation with typical pathology
Can have typical presentation with atypical pathology
Overlap syndromes with varying combinations of ALS, dementia, and parkinsonian features
Atypical presentations that may or may not be a component of the same disease
Asymmetric features at presentation (ALS/PD)
Majority of cases are sporadic
5–10% are familial—tend to be younger and atypical
Probable preclinical syndromes reflecting clinically undetected neurodegeneration
Pathology includes cytoskeletal markers
AD: plaques, tangles, occasional Lewy bodies
PD: Lewy bodies, occasional tangles
ALS: Neurofilamentous inclusions, Bunina bodies, occasional Lewy bodies
Death rate increases with increasing age

AD (Sternberger et al., 1985), and PD (Forno et al., 1984). Lewy bodies have been found in the anterior horn cells of patients with ALS (N. Leigh, *personal communication*; Murayama et al., 1989; McGeer et al., 1988) and in the cortex of patients with dementia (Hansen et al., 1990). Neurofibrillary tangles are found in the cortex and brain stem and Lewy bodies in anterior horn cells of Guamanian patients who present with ALS (Hirano et al., 1961a; D. Perl, *personal communication*). Patients with typical features of PD can have nigral degeneration associated with neurofibrillary tangles but not Lewy bodies (Rajput et al., 1989), and patients with otherwise typical AD may have Lewy bodies and not neurofibrillary tangles in the cortex (Terry et al., 1987). Recent laser microprobe mass analysis (LAMMA) studies further demonstrate a remarkably similar accumulation of the trace metals aluminum and iron in neurofibrillary tangles of patients with AD and ALS/PDC complex of Guam (Good et al., 1993; Perl et al., 1986) and in neuromelanin granules and Lewy bodies in neurons of the substantia nigra of patients with PD (Good et al., 1992).

Conclusion

These observations suggest that the definition of ALS, as with AD and PD, is not straightforward. There exists a clinical and pathological spectrum and individual cases can be seen with motor neuron dysfunction in which there is not universal agreement by clinicians and pathologists as to diagnosis. Outlier and overlap syndromes occur where it is not clear that they represent the same entity that most would call ALS and where the borders with the other neurodegenerative disease are blurred. We are left with these fundamental questions in ALS as we are in the other neurodegenerative disorders: a) What is the basis of the neurodegeneration?, b) Why do some cells die and not others (what is the basis of their specific vulnerability)?, and c) Why do some cells survive and not others (what provides them with their protection)?

References

Bonduelle M (1975): Amyotrophic lateral sclerosis. In: *Handbook of Clinical Neurology, Vol. 22: System Disorders and Atrophy: Part II*, Vinken PJ, Bruyn GW, eds. New York: Elsevier.

Bradley WG (1987): Recent views on amyotrophic lateral sclerosis with emphasis on electro-physiological studies. *Muscle Nerve* 10:490–502.

Bradley WG, Good P, Rasool CG, Adelamn LS (1983): Morphometric and biochemical studies of peripheral nerves in amyotrophic lateral sclerosis. *Ann Neurol* 14:267–277.

Brownell B, Oppenheimer DR, Hughes JT (1970): The central nervous system in motor neurone disease. *J Neurol Neurosurg Psychiatry* 33:338–357.

Buckley J, Warlow C, Smith P, et al. (1983): Motor neuron disease in England and Wales, 1959–1979. *J Neurol Neurosurg Psychiatry* 46:197–205.

Calne DB, Eisen A (1989): The relationship between Alzheimer's disease, Parkinson's disease and motor neuron disease. *Can J Neurol Sci* 16:547–550.

Chida K, Sakamaki S, Takasu T (1989): Alteration in autonomic function and cardiovascular regulation in amyotrophic lateral sclerosis. *J Neurol* 236:127–130.

Chio A, Brignolio F, Meineri P, Schiffer D (1987): Phenotypic and genotypic heterogeneity of dominantly inherited amyotrophic lateral sclerosis. *Acta Neurol Scand* 75:277–282.

Chou S-M (1979): Pathognomy of intraneuronal inclusions in ALS. In: *Amyotrophic Lateral Sclerosis*, Tsubaki T, Toyokura Y, eds. Baltimore: University Park Press.

Deapen DM, Henderson BE (1986): A case-control study of amyotrophic lateral sclerosis. *Am J Epidemiol* 123:790–799.

Delisle MB, Carpenter S (1984): Neurofibrillary axonal swellings and amyotrophic lateral sclerosis. *J Neurol Sci* 63:241–250.

Delisle MB, Gorce P, Hauw JJ, et al. (1987): Motor neuron disease, parkinsonism and dementia. *Acta Neuropathol* 75:104–108.

Ditter SM, Mirra SS (1987): Neuropathologic and clinical features of Parkinson's disease in Alzheimer's disease patients. *Neurology* 37:754–760.

Durrleman S, Alperovitch A (1989): Increasing trend of ALS in France and elsewhere: are the changes real? *Neurology* 39:768–773.

Dyck PJ, Stevens JC, Mulder DW, Espinosa RE (1975): Frequency of nerve fiber degeneration of peripheral motor and sensory neurons in amyotrophic lateral sclerosis: morphometry of deep and superficial peroneal nerves. *Neurology* 25:781–785.

Emery AEH, Holloway S (1982): Familial motor neuron diseases. *Adv Neurol* 36:139–145.

Engel WK, Kurland LT, Klatko I (1959): An inherited disease similar to amyotrophic lateral sclerosis with a pattern of posterior column involvement: an intermediate form? *Brain* 82:203–220.

Forno LS, Sternberger LA, Sternberger NH, et al. (1986): Reaction of Lewy bodies with antibodies to phosphorylated and nonphosphorylated neurofilaments. *Neurosci Lett* 64:253–258.

Gajdusek DC, Salazar AM (1982): Amyotrophic lateral sclerosis and parkinsonian syndromes in high incidence among the Auyu and Jakai people of West New Guinea. *Neurology* 32:107–126.

Good PF, Olanow CW, Perl DP (1993): Iron (Fe) and aluminum (Al) accumulate in the neuromelanin granules of the substantia nigra pars compacta (SNC) of idiopathic Parkinson's disease (PD). *Brain Research* 593:343–346.

Good PF, Perl DP, Bierer LM, Schmeidler J (1992): Selective accumulation of aluminum and iron in the neurofibrillary tangles of Alzheimer's disease: a laser microprobe (LAMMA) study. *Ann Neurol* 31:286–292.

Gubbay SS, Kahana E, Zilber N, et al. (1985): Amyotrophic lateral sclerosis: a study of its presentation and prognosis. *J Neurol* 232:295–300.

Hakim AM, Mathieson G (1979): Dementia in Parkinson's disease: a neuropathologic study. *Neurology* 29:1209–1214.

Hammer RP Jr, Tomiyasu U, Scheibel AB (1979): Degeneration of the human Betz cell due to amyotrophic lateral sclerosis. *Exp Neurol* 63:336–346.

Hansen L, Salmon D, Galasko D (1990): The Lewy body variant of Alzheimer's disease. A clinical and pathologic entity. *Neurology* 40:1–8.

Hirano A, Iwata M (1979): Pathology of motor neurons with special reference to amyotrophic lateral sclerosis and related diseases. In: *Amyotrophic Lateral Sclerosis*, Tsubaki T, Toyokura Y, eds. Baltimore: University Park Press.

Hirano A, Kurland KT, Krooth RS, Lessell S (1961a): Parkinsonism-dementia complex, an endemic disease on the island of Guam. I. Clinical features. *Brain* 84:642–661.

Hirano A, Kurland LT, Sayre GP (1967): Familial amyotrophic lateral sclerosis: a subgroup characterized by posterior and spinocerebellar tract involvement and hyaline inclusions in the anterior horn cells. *Arch Neurol* 16:232–242.

Hirano A, Malamud N, Kurland KT (1961b): Parkinsonism-dementia complex, an endemic disease on the island of Guam. II. Pathological features. *Brain* 84:662–679.

Howard JE, McGill KC, Dorfman JL (1988): Age effects on properties of motor unit action potentials: ADEMG analysis. *Ann Neurol* 24:207–213.

Hudson AJ (1981): Amyotrophic lateral sclerosis and its association with dementia, parkinsonism and other neurological disorders: a review. *Brain* 104:217–247.

Hughes JT (1982): Pathology of amyotrophic lateral sclerosis. *Adv Neurol* 36:61–73.

Ishino H, Higashi H, Kuroda S, et al. (1974): Motor nuclear involvement in progressive supranuclear palsy. *J Neurol Sci* 22:235–244.

Iwata M (1987): Clinico-pathological studies of long-survival ALS cases maintained by active life-support measures. *Adv Exp Med Biol* 209:223–225.

Jellinger LK (1987): Pathologic correlates of dementia in Parkinson's disease. *Arch Neurol* 44:690–691.

Kahana E, Zilber N (1984): Changes in the incidence of amyotrophic lateral sclerosis in Israel. *Arch Neurol* 41:157–160.

Kasarkis EJ, Winslow M (1989): When did Lou Gehrig's personal illness begin? *Neurology* 39:1243–1245.

Katagiri T, Kuzirai T, Nihei K, et al. (1988): Immunocytochemical study of Onuf's nucleus in amyotrophic lateral sclerosis. *Jpn J Med* 27:23–28.

Kondo K (1979): Population dynamics of motor neuron disease. In: *Amyotrophic Lateral Sclerosis*, Tsubaki T, Toyokura Y, eds. Baltimore: University Park Press.

Kondo K (1984): Motor neuron disease and Parkinson's disease are not associated with other disorders at autopsy. *Neuroepidemiology* 3:182–194.

Kondo K, Tsubaki T (1977): Changing mortality patterns of motor neuron disease in Japan. *J Neurol Sci* 32:411–424.

Kurland LT, Mulder DW (1955): Epidemiologic investigations of amyotrophic lateral sclerosis. 2. Familial aggregations indicative of dominant inheritance. *Neurology* 5:182–196; 249–268.

Kurland LT, Mulder DW (1988): Recent epidemiologic developments in the context of earlier observations on ALS: geographic isolates in the Marianas and other islands of the Western Pacific. *Excerpta Medica International Congress Series* No. 769, 3–9.

Kurtzke JF, Beebe GW (1980): Epidemiology of amyotrophic lateral sclerosis. 1. A case-control comparison based on ALS deaths. *Neurology* 30:453–462.

Kuzuhara S, Mori H, Izumiyama N, et al. (1988): Lewy bodies are ubiquitinated. A light and electron microscopic immunocytochemical study. *Acta Neuropathol* 345–353.

Lawyer T Jr, Netsky MG (1953): Amyotrophic lateral sclerosis: a clinicoanatomic study of fifty-three cases. *AMA Arch Neurol Psychiatry* 69:171–192.

Leigh PN, Anderton BH, Dodson A, et al. (1988): Ubiquitin deposits in anterior horn cells in motor neurone disease. *Neursci Lett* 93:197–203.

Leone M, Chandra V, Schoenberg BS (1987): Motor neuron disease in the United States, 1971 and 1973–1978: patterns of mortality and associated conditions at the time of death. *Neurology* 37:1339–1343.

Leverenz J, Sumi SM (1986): Parkinson's disease in patients with Alzheimer's disease. *Arch Neurol* 43:662–664.

Li T-M, Alberman E, Swash M (1988): Comparison of sporadic and familial disease amongst 580 cases of motor neuron disease. *J Neurol Neurosurg Psychiatry* 51:778–784.

Lieberman A, Dziatolowski M, Kupersmith M, et al. (1979): Dementia in Parkinson's disease. *Ann Neurol* 6:355–359.

Lilienfeld DE, Chan E, Ehland J, et al. (1989): Rising mortality from motoneurons disease in the USA, 1962–1984. *Lancet* 1:710–713.

Lowe J, Aldridge F, Lennox G, et al. (1989): Inclusion bodies in motor cortex and brainstem of patients with motor neurone disease are detected by immunocytochemical localisation of ubiquitin. *Neurosci Lett* 105:7–13.

Mayeux R, Rosenstein R, Stern Y, et al. (1986): The prevelance and risk of dementia in idiopathic Parkinson's disease. *Ann Neurol* 20:128.

McGeer PL, Itagaki S, Boyes BE, McGeer EG (1988): Reactive microglia are positive for HLA-DR in the substantia nigra of Parkinson's and Alzheimer's disease brains. *Neurology* 38:1285–1291.

Mitsuyama Y (1984): Presenile dementia with motor neuron disease in Japan: clinicopathological review of 26 cases. *J Neurol Neurosurg Psychiatry* 47: 953–959.

Molsa PK, Martolla RJ, Rinne UK (1984): Extrapyramidal signs in Alzheimer's disease. *Neurology* 34:1114–1116.

Mulder DW, Bushek W, Spring E, Karnes J, Dyck PJ (1983): Motor neuron disease (ALS); evaluation of detection thresholds of cutaneous sensation. *Neurology* 1625–1627.

Mulder DW, Howard FM Jr (1976): Patient resistance and prognosis in amyotrophic lateral sclerosis. *Mayo Clin Proc* 51:537–541.

Mulder DW, Kurland LT, Offord KP, Beard CM (1986): Familial adult motor neuron disease; amyotrophic lateral sclerosis. *Neurology* 36:511–517.

Munsat TL, Andres PL, Finison L, Conlon T, Thibodeau L (1988): The natural history of motoneuron loss in amyotrophic lateral sclerosis. *Neurology* 38:409–413.

Murayama S, Ooraway, Mori H, et al. (1989): Immunocytochemical and ultrasound study of Lewy body like hyaline inclusions in familial amyotrophic lateral sclerosis. *Acta Neuropathol* 78:143–152.

Murros K, Fogelholm R (1983): Amyotrophic lateral sclerosis in Middle-Finland: an epidemiological study. *Acta Neurol Scand* 67:41–47.

Neary D, Snowden JS, Mann DMA, et al. (1990): Frontal lobe dementia and motor neuron disease. *J Neurol Neurosurg Psychiatry* 53:23–32.

Norris FH Jr, U KS, Denys EH, Archibald KC, Lebo C (1978): Amyotrophic lateral sclerosis (letter to the editor). *Mayo Clin Proc* 53:544.

Olanow CW, Drayer B (1987): Brain iron: MRI studies in Parkinson's syndrome. In: *Recent Developments in Parkinson's disease*, II, Fahn S, Marsden CD, Calne DB, Goldstein M, eds. MacMillan Healthcare.

Pearce J (1974): The extrapyramidal disorder of Alzheimer's disease. *Eur Neurol* 12:94–103.

Perl DP, Munoz-Garcia D, Good P, Pendlebuary WW (1986): Laser microprobe mass analysis (LAMMA)- a new approach to the study of the association of aluminum and neurofibrillary tangle formation. In: *Alzheimer's and Parkinson's disease*, Fisher A, Hanin I, Lachman C, eds. New York: Plenum Press.

Perry G, Friedman R, Shaw G, Chau V (1987): Ubiquitin is detected in neurofibrillary tangles and senile plaque neurites of Alzheimer disease brains. *Proc Natl Acad Sci USA* 84:3033–3036.

Rajput AH, Uitti RJ, Sudhakar S, Rozdilsky B (1989): Parkinsonism and neurofibrillary tangle pathology in pigmented nuclei. *Ann Neurol* 25:602–606.

Rowland LP (1980): Motor neuron diseases: the clinical syndromes. In: *The Diagnosis and Treatment of Amyotrophic Lateral Sclerosis*, Mulander, ed. Boston: Houghton Mifflin.

Salazar AM, Masters CL, Gajdusek DC, Gibbs CJ (1983): Syndromes of amyotrophic lateral sclerosis and dementia: relation to transmissible Creutzfeldt-Jakob disease. *Ann Neurol* 14:17–26.

Shy GM, Drager GM (1960): A neurological syndrome associated with orthostatic hypotension: a clinical pathologic study. *Arch Neurol* 2:511–527.

Sobue G, Hashizume Y, Yasuda T, et al. (1990): Phosphorylated high molecular weight neurofilament protein in lower motor neurons in amyotrophic lateral sclerosis and other neurodegenerative diseases involving ventral horn cells. *Acta Neuropathol* 79:402–408.

Sobue G, Matsuoka Y, Mukai E, Takayanagi T, Sobue I (1981): Pathology of myelinated fibers in cervical and lumbar ventral spinal roots in amyotrophic lateral sclerosis. *J Neurol Sci* 50:413–421.

Sternberger LA, Sternberger NH, Ulrich J (1985): Aberrant neurofilament phosphorylation in Alzheimer disease. *Proc Natl Acad Sci USA* 82:4274–4276.

Swash M, Schwartz MS (1982): A longitudinal study of changes in motor units in motor neuron disease. *J Neurol Sci* 56:185–197.

Swash M, Schwartz MS (1984): Staging motor neurone disease: single fibre EMG studies of asymmetry progression and compensatory reinnervation. In: *Research Progress in Motor Neurone Disease*, Rose FC, ed. London: Pitman Books.

Tanaka J, Nakamura H, Tabuchi Y, Takahashi K (1984): Familial amyotrophic lateral sclerosis: features of multisystem degeneration. *Acta Neuropathol (Berlin)* 64:22–29.

Tandan R, Bradley WG (1985): Amyotrophic lateral sclerosis. Part I. Clinical features, pathology, and ethical issues in management. *Ann Neurol* 18:271–280.

Terry RD, Hansen LA, DeTeresa R, et al. (1987): Senile dementia of the Alzheimer type without neurocortical neurofibrillary tangles. *J Neuropathol Exp Neurol* 46:262–268.

Tsukagoshi H, Yin Q, Yamada M, et al. (1988): Morphometric study on the spinal cord of amyotrophic lateral sclerosis: quantification of the number and size of anterior horn cells of the cervical cord. *Excerpta Medica International Congress Series* No. 769, 119–124.

Vejjajiva A, Foster JB, Miller H (1967): Motor neuron disease: a clinical study. *J Neurol Sci* 4:299–314.

Williams DB, Windebank AJ (1991): Motor neuron disease (amyotrophic lateral sclerosis). *Mayo Clin Proc* 66:54–82.

World Federation of Neurology Research Committee Research Group on Neuromuscular Diseases (1988). Classification of neuromuscular diseases. *J Neurol Sci* 86:333–360.

Yoshida S, Mulder DW, Kurland LT, Chu C-P, Okazaki H (1986): Follow-up study on amyotrophic lateral sclerosis in Rochester, Minn., 1925 through 1984. *Neuroepidemiology* 5:61–70.

Younger DS, Chou S, Hays AP, et al. (1988): Primary lateral sclerosis: a clinical diagnosis reemerges. *Arch Neurol* 45:1304–1307.

Chapter 13

Pathogenic Mechanisms in Amyotrophic Lateral Sclerosis and Other Motor Neuron Disorders

P. N. Leigh

Amyotrophic lateral sclerosis (ALS), or motor neuron disease (MND), is a progressive, fatal disorder typified by degeneration of the upper and lower motor neurons in the brain and spinal cord, and corresponding upper and lower motor neuron signs. The diagnosis is based on these signs, the absence of other disease, and evidence of progression (Swash and Leigh, 1992). A brief classification of motor neuron disorders is given in Table 1.

The pathogenesis of ALS and other motor neuron disorders has been reviewed by Tandan and Bradley (1985a,b), by Rowland (1991), and by Leigh (1990, 1991). The field continues to progress rapidly and current themes are reviewed in this chapter.

The Pathogenesis of ALS: Epidemiology and Risk Factors

ALS has a worldwide incidence of around 1 to 2 per 100,000 and a prevalence of about 3 to 5 per 100,000. In the Western Pacific (the Marianas chain of islands including Guam, the Kii peninsula of Japan, and Western New Guinea), incidence and prevalence rates may reach 100 times those in Europe and the USA (Gajdusek, 1982). Age-specific mortality rates for sporadic ALS peak between 50 and 70 years (Kurtzke, 1991). The peak occurs about a decade earlier in familial ALS (Kurland and Mulder, 1955; Emery and Holloway, 1982; Williams et al., 1988). The sex ratio

ADVANCES IN RESEARCH ON NEURODEGENERATION, I
D.B. Calne et al.

in sporadic ALS is 1.5–2:1 (Kurtzke, 1991). Five to 10% of all cases are familial, usually with an autosomal dominant pattern of inheritance (Kurland and Mulder, 1955; Williams et al., 1988; Veltema et al., 1990; Siddique et al., 1991a,b). ALS may be increasing in incidence especially in older age groups (Durrleman and Alperovitch, 1989; Lilienfeld et al., 1989; Kurtzke, 1991). Age, sex, and possibly previous trauma are risk factors for ALS (Kurtzke, 1991). Farming, manual labor, and strenuous physical activity have been implicated as risk factors but although local variations in distribution of ALS cases or death rates have been described, the evidence for geographic clustering is unconvincing (Kurtzke, 1982; Sienko et al., 1990; Gunnarsson et al., 1990; Kurtzke, 1991).

Pathology and Molecular Pathology of ALS

Degeneration in ALS affects somatic motor neurons of the brain stem and spinal cord but typically spares the oculomotor nuclei and Onuf's nucleus in the sacral spinal cord. In the cortex, most cases show degeneration of large pyramidal neurons of the motor cortex (Lawyer and Netsky, 1953; Brownell et al., 1970), although degenerating fibers may extend into prefrontal, parietal, and temporal areas, and into the subthalamic nucleus, substantia nigra, superior colliculus, and brain stem reticular formation (Smith, 1960; Hughes, 1982). Degeneration of Clarke's column is common in ALS (Okamoto et al., 1988; Williams et al., 1990), and patients with prolonged survival may show clinical evidence of supranuclear ophthalmoplegia, and pathological changes of the third nerve nucleus, the red nucleus, substantia nigra, globus pallidus, subthalamus, thalamus, dentate nucleus, brain stem reticular formation, pontine tegmentum, and Clarke's column, with loss of fibers in the corticospinal tracts, the medial lemniscus, and central tegmental tract (Hayashi and Kato, 1989). There is evidence that degeneration of the posterior columns may be more common in familial than in sporadic ALS (Emery and Holloway, 1982).

Bunina bodies (1 to 4-μg eosinophilic cytoplasmic inclusions) are found in 70% to 80% of familial and sporadic ALS cases. Dense bodies, similar or identical to Lewy bodies, occur in about 20% of cases (Bunina, 1962; Chou, 1979; Leigh et al., 1991). Antibodies against ubiquitin reveal a population of inclusions that are more abundant than Bunina bodies and may be specific for ALS (Leigh et al., 1988; Lowe et al., 1988; Murayama et al., 1990a; Mighelli et al., 1990; Leigh et al., 1991; Schiffer et al., 1991; Mizusawa et al., 1991). These take the form either of "skeins," which are bundles or threads of ubiquitin-immunoreactive

Table 1. Human Motor Neuron Disorders

Associated with specific mechanisms
Congenital and familial
- Autosomal recessive inheritance
  - Spinal muscular atrophy (SMA)
    - type 1 (acute infantile)
    - type 2 (intermediate)
    - type 3 (juvenile chronic)
    - type 4 (adult onset, chronic)
    - scapuloperoneal forms
- Hexosaminidase deficiency
- Fazio-Londe disease
- North African and other forms of juvenile onset ALS
- Brown-Vialetto-Van Laere syndrome (juvenile onset pseudobulbar palsy with sensorineural deafness)
- Neuraxonal dystrophy
- X-linked
  - Bulbospinal neuronopathy (Kennedy's syndrome)
- Autosomal dominant
  - Familial ALS
  - Familial ALS with dementia of frontal lobe type

Infective
- Poliomyelitis
- HIV
- ?Spongiform encephalopathies ("prion" disorders)
- Syphilis

Associated with neurotoxins
- Lead-induced motor neuropathy
- Other metals (Mn, Hg, Se, ?Al)
- Neurolathyrism (related to ingestion of chickling pea lathyrus saliva: toxin probably B-oxalyl-amino L-alanine)
- "Mantakassa," "Konzo," and related acute upper motor neurone syndromes (related to ingestion of untreated casave: toxin—probably cyanide)

Metabolic
- Hyperparathyroidism

Autoimmune
- Motor neuron disease/motor neuronopathy with antiganglioside antibodies
- Motor neuron disease/motor neuronopathy with paraprotenemia
- Motor neuron disease with lymphoma

Ischemic myelopathy (radiation-induced)
- Radiation-induced plexopathy

Idiopathic
- Sporadic ALS
- Western Pacific ALS-PD-dementia syndrome
- Monomelic motor neuron disease
- Motor neuron disease with dementia of frontal lobe type
- "Madras" form of motor neuron disease

Amyotrophy in multisystem disorders
- Spinocerebellar degenerations
- Pick's disease
- Parkinson's disease
- Joseph disease
- Huntington's disease
- Spongiform encephalopathies

material, or dense bodies, some of which correspond to Lewy body–like inclusions (Leigh et al., 1988; Murayama et al., 1989). Typical brain stem Lewy bodies are rare in ALS, although diffuse Lewy body disease may be associated with nigral Lewy bodies and anterior horn cell degeneration without spinal Lewy body–like inclusions (Delisle et al., 1990). Ubiquitin-immunoreactive inclusions of the kind seen in sporadic and familial ALS are present in spinal motor neurons in Western Pacific (Guam) ALS (Matsumoto et al., 1990).

Ubiquitinated inclusions may represent an attempt of a damaged cell to rid itself of altered or abnormal proteins via the ubiquitin proteolytic pathway (Rechsteiner, 1987). Ubiquitin is a nonspecific marker for neurodegeneration, although the morphology of the inclusions in ALS is characteristic. Ubiquitin in ALS inclusions has not been associated with other altered neuronal proteins (Leigh et al., 1989, 1991). Ubiquitin-immunoreactive inclusions are found in hippocampal dentate granule cells and in the superficial layers of the frontal and temporal neucortex in patients with ALS and dementia of frontal lobe type (DFLT) (Wightman et al., 1992). Okamoto et al. (1991) and Wightman et al. (1992) found identical inclusions in a few ALS patients who did not show significant cognitive deficits.

Pick's Disease, Dementia of Frontal Lobe Type, and ALS

The pathology of ALS with dementia has been well characterized (Mitsuyama and Takamatsu, 1971; Mitsuyama and Takamiya, 1979; Mitsuyama, 1984; Morita et al., 1987; Neary et al., 1990; Wightman et al., 1992). In general, the motor system pathology is the same in typical ALS and ALS with dementia; ubiquitin-immunoreactive inclusions are present in brain stem and spinal motor neurons in both syndromes (Wightman et al., 1992). Hippocampal and neocortical ubiquitin-immunoreactive inclusions in ALS with dementia differ from Pick bodies in their molecular composition. The inclusions in ALS are nonargyrophilic and are not identified by antibodies against tau and neurofilament proteins (Okamoto et al., 1991; Wightman et al., 1992). Ballooned neurons, which are a common feature of Pick's disease, are not seen in ALS with dementia. Thus, ALS with DFLT can be differentiated from typical Pick's disease on pathological grounds. The dementia syndrome accompanying frontotemporal lobar atrophy is common to cases with Pick bodies and ballooned neurons (typical Pick's disease: Type A of Constantinidis, 1974), and to those with "nonspecific" features, with or without ALS (Kim et al., 1981;

Brun, 1987; Gustaffson, 1987; Neary et al., 1988, 1990). Occasionally, however, ALS and dementia are associated with Pick bodies (Sam et al., 1991).

ALS and DFLT may be familial, with inheritance suggesting an autosomal dominant transmission (Kim et al., 1981; Gunnarsson et al., 1991) but with variable phenotypic expression; some individuals in a kindred presenting with either dementia or ALS (Gunnarsson et al., 1991).

Spongiform Encephalopathy and ALS Dementia

Transmission experiments in patients with the amyotrophy and CJD have been unsuccessful (Salazar et al., 1983; Brown et al., 1984), with one well documented exception. Inoculation of brain material into a squirrel monkey was reported as negative initially but when the animal died without neurological abnormalities after 7 years there was evidence of spongiform degeneration (Allen et al., 1971; Salazar et al., 1983; Connolly et al., 1988). Screening for prion gene mutations in patients with typical sporadic and familial ALS thus far has been negative.

Molecular Genetics

SMA types 1, 2, and 3 have been linked to markers on chromosome 5q12-ql3 (Melki et al., 1990a,b; Brzustowicz et al., 1990; Gilliam et al., 1990). Hexosaminidase deficiencies may mimic SMA (Neote et al., 1991), but the SMA locus is distinct from the Hex-B gene on chromosome 5q. A mutation of the androgen receptor (AR) gene has been identified in X-linked bulbospinal neuronopathy, or Kennedy's syndrome (Harding et al., 1982; Sobue et al., 1989; La Spada et al., 1991). The mutation consists of an enlarged CAG repeat (coding for a polyglutamine tract) in the first exon of the AR gene. It is not known how the abnormally long polyglutamine repeat affects AR gene function, but it may alter transcriptional regulation (La Spada et al., 1991).

ARs are expressed in lower motor neurons but the relationship between AR function and motor neuron survival is unknown. Androgens modify sexually dimorphic neurons such as those that innervate the bulbocavernosus in the rat (Araki et al., 1991). It seems likely that androgens mediate trophic influences between muscle and nerve but these have yet to be defined.

A locus on the long arm of chromosome 21 has been identified in familial ALS (Siddique et al., 1991a,b), but there is evidence of genetic heterogeneity (Siddique et al., 1991a; Conneally, 1991). Most affected individuals show the classical features of ALS, but some individuals within a family with typical ALS may show only lower motor neuron signs (Siddique et al., 1989; Veltema et al., 1990). Indeed, pedigrees have been identified in which ALS and SMA co-exist (Appelbaum et al., 1992). This could be a coincidental association, although this seems unlikely. Dementia with or without parkinsoniam may be more common in familial ALS than in sporadic ALS (Hudson, 1981). Several forms of early-onset progressive spastic paraplegia with muscle wasting ("juvenile ALS") has been reported from North Africa (Ben Hamida et al., 1990). The signs usually are symmetrical, with prolonged survival, and on clinical grounds these syndromes are different from typical ALS. The pathology is unknown.

Excitotoxins

It has long been thought that a toxin in flour from the cycad cycas circinalis may cause the Guam ALS-parkinsonism-dementia complex (PDC). An excitatory amino acid, β-methyl-amino-l-alanine (BMAA), is unlikely to be responsible, but cycasin has not been definitely excluded (Garruto et al., 1988; Duncan et al., 1990; Zhang et al., 1990). Endogenous excitotoxins have been implicated following the observation of Plaitakis and Caroscio (1987) that ALS patients show impaired clearance of serum glutamate after an oral glutamate load. Brain and spinal cord levels of glutamate are diminished, even in areas of the brain not known to be affected by the disease process (Perry et al., 1987; Plaitakis et al., 1988). Motor neuron degeneration occasionally is associated with multisystem disorders such as olivopontocerebellar atrophy (OPCA) in which the activity of glutamate dehydrogenase (GDH) is decreased (Plaitakis et al., 1982, 1984), leading Hugon et al. (1989) to assay GDH activity in leukocytes of ALS patients. A subset of ALS patients had decreased leukocyte GDH activity, although this has not been a consistent finding (Plaitakis and Caroscio, 1987; Aubby et al., 1988). GDH activity in the spinal cord is increased both in the white and gray matter (Malessa et al., 1991), suggesting a generalized abnormality of glutamate metabolism, although these changes may be secondary to gliosis. Cerebrospinal fluid (CSF) from ALS patients has been found to contain elevated (Rothstein et al., 1991) or normal (Perry et al., 1990) levels of glutamate and aspartate, but it is difficult to interpret these changes in terms of the glutamate pool, which is located both in glia and neurons and only part of which is concerned with neurotransmitter glutamate (Plaitakis, 1990). Changes in CSF, brain, and spinal cord glutamate may reflect an increased release of synaptic glutamate or decreased re-uptake of glutamate, the consequences of neuronal loss and/or gliosis, or a mixture of these factors.

Alterations of glycine may contribute to neuronal damage, perhaps through the action of glycine at the allosteric site on the N-methyl-D-aspartate (NMDA) receptor (Plaitakis, 1990; de Belleroche et al., 1984, 1990). Sulfur-containing amino acids such as L-cysteine have been implicated as toxins in ALS but impairment of sulfur oxidation is not specific for ALS (Olney et al., 1990; Heafield et al., 1990).

Pathogenesis of ALS: The Role of Immunological Mechanisms

About 5% of patients with ALS, in contrast to about 2% of controls, have a monoclonal gammopathy (Shy et al., 1986). Lymphoma also may be associated with ALS (Latov, 1982; Younger et al., 1989, 1990; Rowland, 1991). Fifty percent of patients with ALS and lymphoma have upper motor neuron signs (Younger et al., 1989; Rowland, 1991) and CSF protein may be raised with oligoclonal bands in the CSF. A link between other neoplasms and ALS is unproven, but rarely ALS-like syndromes improve after treatment of nonlymphoid tumours (Buchanan and Malamud, 1973; Mitchell and Olczak, 1979; Rosenfeld and Posner, 1991).

Antibodies directed against gangliosides, particularly those containing the Gal-β1-3 GalNAc moiety, such as GM_1, GD_{1b}, and asialo-GM_1, are common in peripheral neuropathy and in syndromes of lower motor neuron degeneration with neurophysiological evidence of multifocal conduction block. The relevance of antiganglioside antibodies in typical ALS is controversial (Shy et al., 1989; Pestronk et al., 1989), since antiganglioside antibodies are usually present in low titer. However, high titers of antiganglioside antibodies directed against GM_1 and other glycoconjugates have been detected in a few patients with upper and lower motor neuron signs (Hays et al., 1990; Sadiq et al., 1990). Patients with antibodies directed against GM_1 and other gangliosides containing the Gal(β1-3) GalNAc moiety may improve with prednisone or cyclophosphamide treatment even if multifocal motor conduction block is not present (Salazar-Grueso et al., 1990; Shy et al., 1990).

It is not known how antiganglioside antibodies might lead to motor neuron degeneration, if indeed they are pathogenic. They may bind either to the surface of motor neurons or to axons in the region of the nodes of Ranvier and paranodal myelin, in a manner similar to that of cholera toxin (Thomas et al., 1990; Latov, 1990), thus disrupting ion channels and precipitating neuronal degeneration. Antibodies could access the perikaryon via the motor endplate, or could interact with growth factor receptors, major histocompatibility (MHC) class 2 molecules, or cell adhesion molecules (Latov, 1990). Mononuclear infiltrates are present in the spinal cord of about 80% of ALS patients (Troost et al., 1989, 1990), and there is increased expression of type 1 and type 2 MHC antigen in macrophages in the white and gray matter of the spinal cord (Lampson et al., 1990; McGeer et al., 1988).

Finally, immunization of guinea pigs with purified preparations of bovine spinal cord motor neurons produces rapidly progressive hindlimb weakness—experimental allergic motor neuron disease (EAMND) (Engelhardt et al., 1989). Upper and lower motor neuron degeneration follows immunization with homogenates of bovine anterior horn gray matter [experimental allergic gray matter disease (EAGMD)] (Engelhardt et al., 1990). The immunological abnormalities in EAMND are associated with damage to the presynaptic terminals of motor neurons, perhaps through damage to calcium channels.

In summary, there is evidence that immunological mechanisms are disturbed in ALS, but there is no evidence to support the use of immunosuppressive treatment in patients with typical ALS. In rare patients with progressive lower motor neuron syndromes, with or without multiple conduction block, and with high titers of polyclonal antiganglioside antibodies or paraproteinemia, immunosuppression may be beneficial.

Viruses and ALS

There is no positive evidence for a viral etiology of ALS (Kennedy, 1990). Probably most interest has focused on polio virus (Miller, 1981; Martyn et al., 1988) in view of its neurotropism, selectivity for lower motor neurons, and the possibility that postpoliomyelitis progressive muscular atrophy (PMMA) might be related to chronic persistent infection. This syndrome is in fact distinct from ALS and comprises (a) a history of acute paralytic poliomyelitis in childhood or adolescence, (2) partial recovery of motor function on functional stability or recovery for at least 15 years, (c) residual muscle atrophy, weakness, and areflexia in at least one limb, but with normal sensation, and (d) development of neuromuscular symptoms and PMMA (Dalakas and Illa, 1991).

Long-term follow-up of cohorts of patients with paralytic poliomyelitis reveal no evidence of an increased risk for ALS over a 20 to 30-year period (Howard et al., 1988; Codd et al., 1985; Armon et al., 1990a); although some patients with antecedent paralytic poliomyelitis have developed ALS (Poskanser et al., 1969; Armon et al., 1990a,b), some patients with slowly progressive disease probably had PPMA (Zilkha, 1962; Campbell et al., 1969; Mulder et al., 1972; Dalakas et al., 1986).

Retroviruses are potential pathogenic agents in motor neuron disorders, although there is no evidence that they can be implicated in ALS (Jubelt, 1991). Several virally induced forms of neuronal degeneration have been investigated (Paquette et al., 1989; Jolicoeur et al., 1991; Contac et al., 1989).

Neurotrophic Substances

There is increasing evidence that neurotrophic factors influence the survival of spinal neurons in tissue culture (Askanas et al., 1991). A neurotrophic factor from muscle has been termed CDF (CAT development factor). CDF stimulates the development of choline acetyl transferase (CAT) in cultures of embryonic rat spinal cord (McManaman et al., 1988) and is identical with leukemia inhibitory factor (LIF). Ciliary neurotrophic factor (CNTF), originally identified in avian ocular tissue (Adler et al., 1979), influences the survival of developing spinal motor neurons and rescues motor neurons from the effects of axotomy (Sendtner et al., 1990) and from naturally occurring cell death in the chick embryo (Oppenheim et al., 1991). CNTF can modify the progression of weakness in mouse models of neurodegeneration (Messer et al., 1987; Mazurkiewicz, 1990; Lindsay et al., 1991; Sendtner et al., 1992). Thyrotropin-releasing hormone (TRH), substance P and bombesin, insulin, insulin-like growth factors, dihydrotestosterone, glucocorticoids, and fibroblast growth factor also influence motor neuron survival (Askanas et al., 1991). Brain-derived neurotrophic factor (BDNF) also may play a role in supporting motor neuron survival after injury. Therapeutic trials of neurotrophic agents in ALS and other motor neuron disorders are planned and proffer hope for rational treatment, perhaps in combination with drugs acting on excitatory amino acid neurotransmission.

Summary

There have been important advances that eventually may reveal the pathogenic mechanisms in ALS and lead to the introduction of treatments that will slow the progression of the disease. The emphasis in future research into the pathogenesis of ALS is likely to turn on molecular genetics, cellular neurobiology, excitotoxicity, and neurotrophic factors. The relevance of autoimmune disturbances remains uncertain, although by no means discounted.

Acknowledgments. The author is grateful to the Motor Neuron Disease Association, the Wellcome Trust, and the Medical Research Council for support. Thanks are due to Mrs. Lesley Gibson for help in preparing the manuscript and to colleagues in the MND research group for their helpful comments.

References

Adler R, Landa KB, Manthorpe M, Varon S (1979): Cholinergic neuronotrophic factors: intraocular distribution of trophic activity for ciliary neurons. *Science* 204:1434–1436.

Allen IV, Dermott E, Connolly JH, Hurwitz LJ (1971): A study of a patient with the amyotrophic form of Creutzfeldt-Jakob disease. *Brain* 94:715–724.

Appelbaum JS, Roos RP, Salazar-Grueso EF, Buchman A, Iannaccone S, Glantz R, Siddique T, Maselli R (1992): Intrafamilial heterogeneity in hereditary motor neuron disease. *Neurology* 42:1488–1492.

Armon C, Daube JR, Windebank AJ, Kurland LT (1990a): How frequently does classic amyotrophic lateral sclerosis develop in survivors of poliomyelitis? *Neurology* 40:172–174.

Armon C, Windebank AJ, Kurland LT, Daube JR (1990b): Polio and ALS. *Neurology* 40:1150–1151.

Askanas V, Mariotti C, Engel WK (1991): Trophic factors for cultured lower motor neurons. In: *Amytrophic Lateral Sclerosis*, Vol 56, Rowland LP, ed. New York: Raven Press.

Aubby D, Saggu HK, Jenner P, Quinn NP, Harding AE, Marsden CD (1988): Leucocyte glutamate dehydrogenase activity in patients with degenerative activity in patients with degenerative neurological disorders. *J Neurol Neurosurg Psychiatry* 51:839–902.

Ben Hamida M, Hentati F, Ben Hamida C (1990): Hereditary motor system diseases (chronic juvenile amyotrophic lateral sclerosis). *Brain* 113:347–363.

Brown P, Rodgers-Johnson P, Cathala F, Gibbs CJ, Gajdusek DC (1984): Creutzfeldt-Jakob disease of long duration: clinicopathological characteristics, transmissability, and differential diagnosis. *Ann Neurol* 16:295–304.

Brownell B, Oppenheimer DR, Trevor Hughes J (1970): The central nervous system in motor neurone disease. *J Neurol Neurosurg Psychiatry* 33:338–357.

Brun A (1987): Frontal lobe degeneration of non-Alzheimer type. *Arch Gerontol Geriatr* 6:193–208.

Brzustowicz LM, Lehner T, Castilla LH, Penchaszadeh GK, Wilhelmsen KC, Daniels R, Davies KE, Leppert M, Ziter F, Wood D, Dubowitz V, Zerres K, Hausmanowa-Petrusewicz I, Ott J, Munsat TL, Gilliam TC (1990): Genetic mapping of chronic childhood-onset spinal muscular atrophy to chromosome 5q11.2–13.3. *Nature* 344:540–541.

Buchanan DS, Malamud N (1973): Motor neuron disease with renal cell carcinoma and post-operative neurologic remission: a clinicopathological report. *Neurology* 23:891–894.

Bunina TL (1962): On intracellular inclusions in familial amyotrophic lateral sclerosis. *Zhurnal Nevropathologii i Psikhiatrii iment SS Korsakova* 62:1293–1299 (in Russian).

Campbell AMG, Williams ER, Pearce J (1969): Late motor neurone degeneration following poliomyelitis. *Neurology* 19:1101–1106.

Charcot JM, Joffroy A (1869): Deux cas d'atrophie musculaire progressive avec les lésions de la substance grise et faisceraux antérolatéaux de la moëlle épinière. *Arch Physiol Normal Patholog* Paris 2:354–367, 629–649, 744–760.

Chou SM (1979): Pathognomy of intraneuronal inclusions in ALS. In: *Amyotrophic Lateral Sclerosis*, Tsubaki T, Toyokura Y, eds. Tokyo: Tokyo University Press; Baltimore: University Park Press.

Codd MB, Mulder DW, Kurland LT, Beard CM, O'Fallon WM (1985): Poliomyelitis in Rochester, Minnesota, 1935–1955: epidemiology and long-term sequelae: a preliminary report. In: *Late Effects of Poliomyelitis*, Halstead LS, Wiechers DO, eds. Miami: Symposia Foundation.

Conneally PM (1991): A first step toward a molecular genetic analysis of amyotrophic lateral sclerosis. *N Engl J Med* 324(20):1430–1432.

Connolly JH, Allen IV, Dermott E (1988): Transmissible agent in the amyotrophic form of Creutzfeldt-Jakob disease. *J Neurol Neurosurg Psychiatry* 51:1459–1460.

Constantinidis J, Richard J, Tissot R (1974): Pick's disease: histological and clinical correlations. *Eur Neurol* 11:208–217.

Contac CH, Harty JT, Plagemann GW (1989): Dual virus etiology of age-dependent poliomyelitis of mice: a potential model for human motor neuron disease. *Microbiol Pathogen* 6:391–401.

Dalakas MC, Elder G, Hallett M, et al. (1986): A long-term follow-up study of patients with post-poliomyelitis neuromuscular symptoms. *N Engl J Med* 314:959–963.

Dalakas MC, Illa I (1991): Post-polio syndrome: concepts in clinical diagnosis, pathogenesis and etiology. In: *Amyotrophic Lateral Sclerosis, Vol 56, Advances in Neurology*, Rowland LP, ed. New York: Raven Press.

de Belleroche J, Recordati A, Rose CF (1984): Elevated levels of amino acids in the CSF of motor neuron disease. *Neurochem Pathol* 2:1–6.

de Belleroche J, Lane RJM, Bandopadhyay R, Rose CF (1990): Abnormalities in amino acid metabolism in ALS. In: *Amyotrophic Lateral Sclerosis. New Advances in Neurotoxicology and Epidemiology*, Rose CF, Norris FH, eds. London/San Francisco: Smith Gordon/Nishimura.

Delisle MB, Gorce P, Hirsch E, Hauw JJ, Rascol A, Bouissou H (1987): Motor neuron disease: parkinsonism and dementia: a report of case with diffuse Lewy-like inclusions. *Acta Neuropathol* 75:104–108.

Duncan MW, Steele JC, Kopin IJ, Markey SP (1990): 2-Amino-3-(methylamino)-propanoic acid (BMAA) in cycad flour. *Neurology* 40:7647–772.

Durrleman S, Alperovitch A (1989): Increasing trend in ALS in France and elsewhere: are the changes real? *Neurology* 39:768–773.

Emery AEH, Holloway S (1982): Familial motor neuron diseases. In: *Human Motor Neuron Diseases*, Rowland LP, ed. New York: Raven Press.

Engelhardt JI, Appel SH, Killian JM (1989): Experimental autoimmune motor neuron disease. *Ann Neurol* 26:368–376.

Engelhardt JI, Appel SH, Killian JM (1990): Motor neuron destruction in guinea pigs immunized with bovine spinal and ventral horn homogenate: experimental autoimmune grey matter disease. *J Neuroimmunol* 27:21–31.

Gajdusek DC (1982): Foci of motor neuron disease in high incidence populations of East Asia and the Western Pacific. In: *Human Motor Neuron Diseases*, Rowland LP, ed. New York: Raven Press.

Garcia J, Engelhardt JI, Appel SH, Stefani E (1990): Increased MEPP frequency as an early sign of experimental immune-mediated motor neuron disease. *Ann Neurol* 28:329–334.

Garruto RM, Yanagihara R, Gajdusek DC (1988): Cycads and amyotrophic lateral sclerosis/parkinsonism dementia. [Letter] *Lancet* 1079.

Gilliam TC, Brzustowicz LM, Castilla LH, Lehner T, Penchaszadeh GK, Daniels RJ, Byth BC, Knowles J, Hislop JE, Shapira Y, Dubowitz V, Munsat TL, Ott J, Davies KE (1990): Genetic homogeneity between acute and chronic forms of spinal muscular atrophy. [Letter] *Nature* 345:823–825.

Gunnarsson L-G, Dahlborn K, Strandman E (1991): Motor neuron disease and dementia reported among 13 members of a single family. *Acta Neurol Scand.*

Gunnarsson L-G, Lindberg G, Söderfelt B, Axelson O (1990): The mortality of motor neuron disease in Sweden. *Arch Neurol* 47:42–46.

Gustafson L (1987): Frontal lobe degeneration of non-Alzheimer type. II. Clinical picture and differential diagnosis. *Arch Gerontol Geriatr* 6:209–223.

Harding AE, Thomas PK, Baraitser M, Bradbury PC, Morgan-Hughes JA, Poneford JR (1982): X-linked recessive bulbospinal neuronopathy: a report of ten cases. *J Neurol Neurosurg Psychiatry* 45:1012–1019.

Hayashi H, Kato S (1989): Total manifestations of amyotrophic lateral sclerosis (ALS) in the totally locked-in state. *J Neurol Sci* 93:19–35.

Hays AP, Roxas A, Sadiq SA, Vallejos H, D'Agati V, Thomas FP, Torres R, Sherman WH, Bailey-Braxton D, Hayes AG, Rowland LP, Latov N (1990): A monoclonal IgA in a patient with amyotrophic lateral sclerosis reacts with neurofilaments and surface antigen on neuroblastoma cells. *J Neuropathol Exp Neurol* 49(4):383–398.

Heafield MT, Fearn S, Steventon GB, Waring RH, Williams AC, Sturman SG (1990): Plasma cysteine and sulfate levels in patients with motor neurone, Parkinson's, and Alzheimer's disease. *Neurosci Lett* 110:216–220.

Howard RS, Wiles CM, Spencer GT (1988): The late sequelae of poliomyelitis. *Q J Med* 66:219–232.

Hughes JT (1982): Pathology of amyotrophic lateral sclerosis. In: *Human Motor Neuron Disease*, Rowland LP, ed. New York: Raven Press.

Hugon J, Tabaraud F, Rigaud M, Vallat J-M, Dumas M (1989): Glutamate dehydrogenase and aspartate aminotransferase in leukocytes of patients with motor neuron disease. *Neurology* 39:956–958.

Jolicoeur P, Passant E, DesGroseillers L, Robibaille Y, Paquette Y, Kay DG (1991): *Amyotrophic Lateral Sclerosis and Other Motor Neurone Diseases, Vol 56, Advances in Neurology*, Rowland LP, ed. New York: Raven Press.

Jubelt B (1991): Viruses and motor neuron diseases. *Amyotrophic Lateral Sclerosis and Other Motor Neurone Diseases, Vol 56, Advances in Neurology*, Rowland LP, ed. New York: Raven Press.

Kennedy PGE (1990): On the possible role of viruses in the aetiology of motor neuron disease: a review. *J Roy Soc Med* 83:784–787.

Kim RC, Collins GH, Parisi JE, Wright WA, Young BC (1981): Familial dementia of adult onset with pathological findings of a 'non-specific' nature. *Brain* 104:61–78.

Kurland LT, Mulder DW (1955): Epidemiological investigations of amyotrophic lateral sclerosis. 2. Familial aggregations indicative of dominant inheritance, parts I and II. *Neurology* 5:182–196, 249–269.

Kurtzke JF (1982): Epidemiology of amyotrophic lateral sclerosis. In: *Human Motor Neuron Diseases, Vol 36, Advances in Neurology*, Rowland LP, ed. New York: Raven Press.

Kurtzke JF (1991): Risk factors in amyotrophic lateral sclerosis. In: *Amyotrophic Lateral Sclerosis and Other Motor Neuron Disorders, Vol 56, Advances in Neurology*, Rowland LP, ed. New York: Raven Press.

Lampson LA, Kushner PD, Sobel RA (1990): Major histocompatibility complex antigen expression in the affected tissues in amyotrophic lateral sclerosis. *Ann Neurol* 28:365–372.

La Spada AR, Wilson EM, Lubahn DE, Harding AE, Fischbeck KH (1991): Androgen receptor gene mutations in X-linked spinal and bulbar muscular atropy. *Nature* 352:77–79.

Lassman H, Weiler R, Fischer P, Bancher C, Jellinger K, Floor E, Danielczyk

W, Seitelberger F, Winkler H (1992): Synaptic pathology in Alzheimer's disease: immunological data for markers of synaptic and large dense-core vesicles. *Neuroscience* 46:1–8.

Latov N (1982): Plasma cell dyscrasia and motor neuron disease. In: *Human Motor Neuron Diseases*, Rowland LP, ed. New York: Raven Press.

Latov N (1990): Neuropathy and anti-GM_1 antibodies. *Ann Neurol* 27(suppl): S41–S43.

Lawyer T, Netsky MG (1953): Amyotrophic Lateral Sclerosis. A Clinicoanatomic study of 53 cases. *Arch Neurol Psychiatry* 69:171–192.

Leigh PN (1990): Amyotrophic lateral sclerosis and other motor neurone disorders. *Curr Opinion Neurol Neurosurg* 3:567–575.

Leigh PN (1991): Amyotrophic lateral sclerosis and other motor neurone disorders. *Curr Opinion Neurol Neurosurg* 4:586–596.

Leigh PN, Anderton BH, Dodson A, Gallo J-M, Swash M, Power DM (1988): Ubiquitin deposits in anterior horn cells in motor neuron disease. *Neurosci Lett* 93:197–203.

Leigh PN, Dodson A, Swash M, Brion JP, Anderton BH (1989): Cytoskeletal abnormalities in motor neurone disease. An immunocytochemical study. *Brain* 112:521–535.

Leigh PN, Whitwell H, Garofalo O, Buller J, Swash M, Martyn JE, Gallo J-M, Weller RO, Anderton BH (1991): Ubiquitin-immunoreactive intraneuronal inclusions in amyotrophic lateral sclerosis: morphology, distribution, and specificity. *Brain* 114:775–788.

Lepage M, Froissart M, Galibert P (1975): Poliomyélite antérieure subaigue. Etude clinique. A propos d'un cas avec isolement d'un poliovirus type II dans le LCR. *Rev Neurol (Paris)* 131:721–724.

Lien LY, Boyce FM, Kleyn P, Brzustowicz LM, Menninger J, Ward DC, Gilliam C, Kunkel LM (1991): Mapping of the human microtubule associated protein 1B in proximity to the spinal muscular atrophy locus at 5913. *Proc Natl Acad Sci USA* 88:7873–7876.

Lilienfeld DE, Chan E, Ehland J, Goobold J, Landrigan PJ, Marsh G, Perl DP (1989): Rising mortality from motor neuron disease in the USA, 1962–1984. *Lancet* i:710–713.

Lowe J, Lennox G, Jefferson D, Morrell K, McQuire D, Gray T, Landon M, Mayer RJ (1988): A filamentous inclusion body within anterior horn neurons in motor neuron disease defined by immunocytochemical localization of ubiquitin-protein conjugates. *Neurosci Lett* 94:203–210.

Malessa S, Leigh PN, Bertel O, Sluga E, Hornykiewicz O (1991): Amyotrophic lateral sclerosis: glutamate dehydrogenase and transmitter amino acids in the spinal cord. *J Neurol Neurosurg Psychiatry* 54:984–988.

Martyn CN, Barker DJP, Osmond C (1988): Motor neuron disease and past poliomyelitis in England and Wales. *Lancet* 1:1319–1322.

Matsumoto S, Hirano A, Goto S (1990): Spinal cord neurofibrillary tangles of Guamanian amyotrophic lateral sclerosis and parkinsonism-dementia complex. *Neurology* 40:975–979.

Mazurkiewicz JE (1990): Ubiquitin deposits in spinal motoneurons on the Mnd (motor neuron degeneration) mouse. *J Neurol Sci* 985:349–350.

McGeer PL, Itagaki S, McGeer EG (1988): Expression of the histocompatibility glycoprotein HLA-DR in neurological disease. *Acta Neuropathol* 76:550–557.

McManaman J, Crawford F, Clark R, Richter J, Fuller F (1989): Multiple neurotrophic factors from skeletal muscle: demonstration of effects of basic fibroblast growth factor and comparison with the 22-kilodalton choline acetyl transferase development factor. *J Neurochem* 53:1763–1771.

Melki J, Sheth P, Abdelhak S, Burlet P, Bachelot M-F, Lathrop MG, Frezal J, Munnich A (1990a): Mapping of acute (type 1) spinal muscular atrophy to chromosome 5q12–q14. *Lancet* 336:271–273.

Melki J, Abdelhak S, Sheth P, Bachelot MF, Burlet P, Marcadet A, Aicardi J, Barois A, Carriere JP, Fardeau M, Fontan D, Ponsot G, Billette T, Angelini C, Barbosa C, Ferriere G, Lanzil G, Ottolini A, Babron MC, Cohen D, Hanauer A, Clerget-Darpoux F, Lathrop M, Munnich A, Frezal J (1990b): Gene for chronic proximal spinal muscular atrophies maps to chromosome 5q. *Nature* 344:764–768.

Messer A, Strominger NL, Mazurkiewicz JE (1987): Histopathology of the late onset degeneration (Mnd) mutant in the mouse. *J Neurogenet* 4:201–213.

Mighelli A, Antilio-Gambetti L, Gambetti P, Mocellini C, Vighani MC, Schiffer D (1990): Ubiquitinated inclusions in spinal cord of patients with motor neuron disease. *Neurosci Lett* 114:5–10.

Miller JR (1981): Prolonged intracerebral infection with poliovirus in asymptomatic mice. *Ann Neurol* 9:590–596.

Mitchell M, Okzak A (1979): Remission of a syndrome indistinguishable from motor neurone disease after resection of a bronchial carcinoma. *Br Med J* 2:176–177.

Mitsuyama Y (1988): ^{123}IMP-SPECT finding of a case of presenile dementia with motor neuron disease. Comparison between Alzheimer's disease. Pick's disease and presenile dementia with motor neuron disease. *Psychiatr Neurol Jpn* 5:499–510.

Mitsuyama Y, Takamatsu I (1971): An autopsy case of presenile dementia with motor neuron disease. *Brain Nerve (Tokyo)* 23:409–416 (in Japanese and English abstract).

Mitsuyama Y, Takamiya S (1979): Presenile dementia with motor neuron disease in Japan: a new entity? *Arch Neurol* 36:592–593.

Mizusawa H, Nakamura H, Wakayama I, Yen S-H C, Hirano A (1991): Skein-like inclusions in the anterior horn cells in motor neuron disease. *J Neurol Sci* 105:14–21.

Morita K, Kaiya H, Ikeda T, Namba M (1987): Pre-senile dementia combined with amyotrophy: a review of 34 Japanese cases. *Arch Gerontol Geriatr* 6:263–277.

Mulder DW, Rosenbaum RA, Layton DD Jr (1972): Late progression of poliomyelitis or forme fruste amyotrophic lateral sclerosis? *Mayo Clin Proc* 47:756–761.

Murayama S, Mori H, Ihara Y, Bouldin TW, Suzuki K, Tomonaga M (1990a): Immunocytochemical and ultrastructural studies of lower motor neurons in amyotrophic lateral sclerosis. *Ann Neurol* 27:137–148.

Murayama S, Mori H, Ihara Y, Tomonaga H (1990b): Immunocytochemical and ultrastructural studies of Pick's disease. *Ann Neurol* 27:394–405.

Murayama S, Okawa Y, Mori H, Nakano I, Ihara Y, Kuzuhara S, Tomonaga M (1989): Immunocytochemical and ultrastructural study of Lewy body-like hyaline inclusions in familial amyotrophic lateral sclerosis. *Acta Neuropathol* 78:143–152.

Neary D, Snowden JS, Mann DMA, Northern B, Boulding PJ, MacDermott N (1990): Frontal lobe dementia and motor neuron disease. *J Neurol Neurosurg Psychiatry* 53:23–32.

Neote K, Mahuran DJ, Cravel RA (1991): Molecular genetics of β-hexosaminidase deficiencies. In: *Amyotrophic Lateral Sclerosis, Vol 56, Advances in Neurology*, Rowland LP, ed. New York: Raven Press.

Norris FH, Denys EH, Sang UK (1990): Polio and ALS. [Letter] *Neurology* 40:1150.

Okamoto K, Hirai S, Yamazaki T, Sun X, Nakazato Y (1991): New ubiquitin-positive intraneuronal inclusions in the extra-motor cortices of patients with ALS. *Neurosci Lett* 129:233–236.

Okamoto K, Yamazaki T, Yamaguchi H, Suoji M, Hirai S (1988): Pathology of Clarke's nucleus in sporadic amyotrophic lateral sclerosis. *Clin Neurol* 28:536–542.

Olney JW, Zorumski C, Price MT, Labruyere J (1990): L-cysteine, a bicarbonate-sensitive endogenous excitotoxin. *Science* 248:596–599.

Oppenheim RW, Prevette D, Qin-Wei Y, Collins F, MacDonald J (1991): Control of embryonic motoneuron survival in vivo by ciliary neurotrophic factor. *Science* 251:1616–1618.

Paquette Y, Hanna Z, Savard P, Brousseau R, Robitaille Y, Jolicoeur P (1989): Retrovirus-induced murine motor neuron disease: mapping the determinant of spongiform degeneration within the envelope gene. *Proc Natl Acad Sci USA* 86:3896–3900.

Perry TL, Hanson S, Jones K (1987): Brain glutamate deficiency in ALS. *Neurology* 37:1845–1848.

Perry TL, Krieger C, Hansen S, Eisen A (1990): Amyotrophic lateral sclerosis: amino acid levels in plasma and cerebrospinal fluid. *Ann Neurol* 28:12–17.

Pestronk A, Adams RN, Cornblath D, Kunci RW, Drachman DB, Clawson L (1989): Patterns of serum IgM antibodies to GM_1 and GD_{1a} gangliosides in amyotrophic lateral sclerosis. *Ann Neurol* 25:98–102.

Pestronk A, Chaudry V, Feldman EL, Griffin JW, Cornblath DR, Denys EH, Gaslberg M, Kunch RW, Olney RK, Yee WC (1990): Lower motor neuron syndromes defined by patterns of weakness, nerve conduction abnormalities and high titers of antiglycolipid antibodies. *Ann Neurol* 27:316–326.

Plaitakis A (1990): Glutamate dysfunction and selective motor neuron degeneration in ALS: an hypothesis. *Ann Neurol* 28:3–8.

Plaitakis A, Berl S, Yahr MD (1982): Abnormal glutamate metabolism in an adult-onset degenerative neurological disorder. *Science* 216:193–196.

Plaitakis A, Caroscio JT (1987): Abnormal glutamate metabolism in amyotrophic lateral sclerosis. *Ann Neurol* 22:575–579.

Plaitakis A, Constantakakis E, Smith J (1988): The neuroexcitotoxic amino acids glutamate and aspartate are altered in the spinal cord and brain in amyotrophic lateral sclerosis. *Ann Neurol* 24:446–449.

Plaitakis A, Nicklaus WJ, Desnick RJ (1980): Glutamate dehydrogenase deficiency in three patients with spinocerebellar syndromes. *Ann Neurol* 7:297–303.

Poskanzer DC, Cantor HM, Kaplan GS (1969): The frequency of preceding poliomyelitis in amyotrophic lateral sclerosis. In: *Motor Neuron Diseases: Research on Amyotrophic Lateral Sclerosis and Related Disorders*, Norris FH Jr, Kurland LT, eds. New York: Grune & Stratton.

Rechsteiner M (1987): Ubiquitin-mediated pathways for intracellular proteolysis. *Annu Rev Cell Biol* 3:1–30.

Rosenfeld MR, Posner JB (1991): Paraneoplastic motor neuron disease. In: *Amyotrophic Lateral Sclerosis and Other Motor Neuron Diseases, Vol 56, Advances in Neurology*, Rowland LP, ed. New York: Raven Press.

Rothstein JD, Kuncl R, Chaudhry V, Clawson L, Cornblath DR, Coyle JT, Drachman DB (1991): *Ann Neurol* 30(2):224–225.

Rothstein JD, Tsai G, Kuncl RW, Clawson L, Cornblath DR, Drachman DB, Pestronk A, Stauch BL, Coyle JT (1990): Abnormal excitatory amino acid metabolism in amyotrophic lateral sclerosis. *Ann Neurol* 28:18–25.

Rowland LP, Defendini R, Sherman WH, et al. (1981): Macroglobulinemia with peripheral neuropathy simulating motor neuron disease. *Ann Neurol* 11:532–536.

Rowland LP (1991): Ten central themes in a decade of ALS research. In: *Amyotrophic Lateral Sclerosis, Vol 56, Advances in Neurology*, Rowland LP, ed. New York: Raven Press.

Sadiq SA, Thomas FP, Kilidireas K, Protopsaltis S, Hayes AP, Lee K-W, Romas SN, Kumar N, van den Berg L, Santoro M, Lange DJ, Younger DS, Lovelace RE, Trojaborg W, Sherman WH, Miller JR, Minuk J, Fehr MA, Roelofs RI, Hollander D, Nichols FT, Mitsumoto H, Kelley JJ, Swift TR, Munsat TL, Latov N (1990): The spectrum of neurologic disease associated with anti-GM1 antibodies. *Neurology* 40:1067–1072.

Salazar AM, Masters CL, Gajdusek C, Gibbs CJ (1983): Syndromes of amyotrophic lateral sclerosis and dementia: relation to transmissible Creutzfeldt-Jakob disease. *Ann Neurol* 14:17–26.

Salazar-Grueso EF, Routbort MJ, Martyn J, Dawson G, Roos RP (1990): Polyclonal IgM anti-GM_1 ganglioside antibody in patients with motor neuron disease and variants. *Ann Neurol* 27:558–563.

Sam M, Gertmann L, Schochet SS, Doshi H (1991): Pick's disease: a case clinically resembling amyotrophic lateral sclerosis. *Neurology* 41:1831–1833.

Santoro M, Thomas FP, Fink ME, Lange DJ, Uncini A, Wadia NH, Latov N, Hayes AP (1990): IgM deposits at nodes of Ranvier in a patient with amyotrophic lateral sclerosis, anti-GM_1 antibodies, and multifocal motor conduction block. *Ann Neurol* 28:373–377.

Schiffer D, Autilio-Gambetti L, Chio A, Gambetti P, Giordana MT, Gullotta F, Migheli A, Vigliani MC (1991): Ubiquitin in motor neuron disease: study at the light and electron microscope. *J Neuropathol Exp Neurol* 4:463–473.

Sendtner M, Kreutzberg GW, Thoenen H (1990): Ciliary neurotropic factor prevents the degeneration of motor neurons after axotomy. *Nature* 345:440–441.

Sendtner M, Schmalbruch H, Stockli KA, Carroll P, Kreutzberg GW, Thoenen H (1992): Ciliary neurotrophic factor prevents degeneration of motor neurons in mouse mutant progressive motor neuronopathy. *Nature* 358:502–504.

Sharpe AH, Junter JJ, Chassler P, Jaenisch R (1990): Role of abortive retroviral infection of neurons in spongiform CNS degeneration. *Nature* 346:181–183.

Shy ME, Evans VA, Lublin FD, Knobler RL, Heimann-Patterson J, Tahmoush AJ, Parry G, Schick P, Deryck TG (1989): Antibodies to GM_1 and GD_{1b} in patients with motor neuron disease without plasma cell dyscrasia. *Ann Neurol* 25:511–513.

Shy ME, Heiman-Patterson T, Parry GJ, Tahmoush A, Evans VA, Schick PK (1990): Lower motor neuron disease in a patient with auto-antibodies against Gal(β1-3) GalNAc in gangliosides GM1 and GD1b: improvement following immunotherapy. *Neurology* 40:842–844.

Shy ME, Rowland LP, Smith T, et al. (1986): Motor neuron disease and plasma cell dyscrasia. *Neurology* 36:1429–1436.

Siddique T, Brown RH (1991a): Genetic basis of familial amyotrophic lateral sclerosis. *N Engl J Med* 325:1383.

Siddique T, Figlewica DA, Pericak-Vance MA, Haines JL, Rouleau G, Jeffers AJ, et al. (1991b): Linkage of a gene causing familial amyotrophic lateral sclerosis to chromosome 21 and evidence of genetic-locus heterogeneity. *N Engl J Med* 324:1381–1384.

Siddique T, Pericak-Vance MA, Brook BR, Roos RP, Hung W-Y, Antel JP, Munsat TL, Phillips K, Warner K, Speer M, Bias WB, Siddique NA, Roses AD (1989): Linkage analysis in familial amyotrophic lateral sclerosis. *Neurology* 39:919–926.

Sienko DG, Davis JP, Taylor JA, Brooks BR (1990): Amyotrophic lateral sclerosis. A case-control study following detection of a cluster in a small Wisconsin community. *Arch Neurol* 47:38–41.

Smith MC (1960): Nerve fibre degeneration in the brain in amyotrophic lateral sclerosis. *J Neurol Neurosurg Psychiatry* 23:269–282.

Sobue G, Hashizuma Y, Mukai E, Hirayama M, Mitsuma T, Takahishi A (1989):

X-linked bulbospinal neuronopathy. A clinicopathological study. *Brain* 112: 209–232.

Sobue G, Hashizuma Y, Yasuda T, Mukai E, Kumagai T, Mitsuma T, Trojanowski JQ (1990): Phosphorylated high molecular weight neurofilament protein in lower motor neurons in amyotrophic lateral sclerosis and other neurodegenerative diseases involving ventral horn cells. *Acta Neuropathol* 79:402–408.

Stallones L, Kasarskis EJ, Stipanowich C, Snider G (1989): Secular trends in mortality rates from motor neuron disease in Kentucky, 1964–1984. *Neuroepidemiology* 8:68–78.

Tandan R, Bradley WG (1985a): Amyotrophic lateral sclerosis. 1. Clinical features, pathology, and ethical issues in management. *Ann Neurol* 18:271–280.

Tandan R, Bradley WG (1985b): Amyotrophic lateral sclerosis. II. Etiopathogenesis. *Ann Neurol* 18:419–431.

Thomas FP, Lee AM, Romas SN, Latov N (1989): Monoclonal IgMs with anti-Gal(β1-3) GalNAc activity in lower motor neuron disease: identification of glycoprotein antigens in neural tissue and cross-reactivity with serum immunoglobulins. *J Neuroimmunol* 23:167–174.

Thomas FP, Thomas JE, Sadiq SA, van den Berg LH, Roelofs RI, Latov M, Hayes AP (1990): Human monoclonal IgM anti-Gal(β1-3) GalNAc autoantibodies bind to the surface of bovine spinal motoneurons. *J Neuropathol Exp Neurol* 49(2):89–95.

Troost D. van den Ord JJ, de Jong JMBV (1990): Immunohistochemical characterization of the inflammatory infiltrate in amyotrophic lateral sclerosis. *Neuropathol Appl Neurobiol* 6:401–410.

Troost D, van den Ord JJ, de Jong JMBV, Swaab DF (1989): Lymphocytic infiltration in the spinal cord of patients with amyotrophic lateral sclerosis. *Clin Neuropathol* 8(6):289–294.

Veltema AN, Roos RAC, Bruyn GW (1990): Autosomal dominant adult amyotrophic lateral sclerosis. *J Neurol Sci* 97:93–115.

Wightman G, Anderson VER, Martin J, Swash M, Anderton BH, Neary D, Mann D, Luthert P, Leigh PN (1992): Hippocampal and neocortical ubiquitin-immunoreactive inclusions in amyotrophic lateral sclerosis with dementia. *Neurosci Lett.*

Williams C, Kozlowski MA, Hinton DR, Miller CA (1990): Degeneration of spinocerebellar neurons in amyotrophic lateral sclerosis. *Ann Neurol* 27:215–225.

Williams DB, Floate DA, Leicester J (1988): Familial motor neuron disease: differing penetrance in large pedigrees. *J Neurol Sci* 86:215–230.

World Federation of Neurology Research Committee on Neuromuscular Disease (1988): Classification of neuromuscular diseases. *J Neurol Sci* 86:333–360.

Young AB (1990): What's the excitement about excitatory amino acids in amyotrophic lateral sclerosis? *Ann Neurol* 28:9–11.

Younger DS, Rowland LP, Latov N, Hays AP, Lange DJ, Sherman W, Inghirami

G, Pesce MA, Knowles DM, Powers J, Miller JR, Fetell MR, Lovelace RE (1991): Lymphoma, motor neuron diseases, and amyotrophic lateral sclerosis. *Ann Neurol* 29:78–86.
Younger DS, Rowland LP, Sherman WH, et al.(1989): Lymphoma, motor neuron diseases, and ALS. *Ann Neurol* 26:187.
Younger DS, Rowland LP, Latov N, et al. (1990): Motor neuron disease and ALS: relation of high CSF protein content to paraproteinaemia and clinical syndromes. *Neurology* 595–599.
Zhang Z, Anderson DW, Mantel N (1990): Geographic patterns of parkinsonism-dementia complex in Guam. *Arch Neurol* 47:1069–1074.
Zilkha KJ (1962): Discussion on motor neurone disease. *Proc R Soc Med* 55:1028–1029.

Chapter 14

Xenobiotic Biochemistry of Neurodegenerative Disease with Particular Attention to Methylation Pathways

Adrian Williams

A plausible hypothesis for several diseases associated with increasing age, as well as some phenomena called by some "unsuccessful aging," is that they are caused in part by damage mediated by chemicals. If so, two basic principles of toxicology must be relevant: that is, that most exogenous chemicals are extensively and quickly biotransformed and that there is enormous variation in the population both qualitatively and quantitatively in the metabolism of such xenobiotics. Thus, an important risk factor for the development of chemical damage that may be at least as important as target cell sensitivity may be the type and extent of biotransformation of a given potential toxin (Mountain, 1963). Not that one should see these necessarily as rival theories, as hypersusceptibility to a chemical could happen in series with both an unsafe biotransformation and unusually weak cellular defenses being necessary before disease occurs (Fig. 1).

In addition to cellular defenses, ability to regenerate and compensation by other neurons and the amount of functional reserve in any given neuronal pool, whether genetically determined or surviving from previous insults, is bound to be important. It also may be necessary to see neurodegenerative disease as a multistage process whereby at a given level of damage, at least in some individuals, vicious cycles are generated that cause further damage. Others have mentioned using free radical arguments this scenario as remaining neurons maybe metabolically overworked.

ADVANCES IN RESEARCH ON NEURODEGENERATION, I
D.B. Calne et al.

Immunocompetence in removing the dead and dying cells also may be important. Unwittingly studying such vicious cycles could be misleading in that one might believe one had found the primary defect or a new subclassification of disease. Different individuals may respond in different ways to the same stimulus and may need different approaches to damage limitation. Indeed, the same chemical exposure could lead in some to one neurodegenerative disease, in others to another, and in other individuals to an autoimmune disease or cancer, depending on either cellular responses or differences in biotransformation. For this reason we must expect some nasty jolts if we are hoping our clinical classification is going to translate smoothly into an etiological classification that in these multifactorial diseases will be a complex mixture of external and genetically determined risk factors.

Ecogenetics

In 1957 Motulsky found that specific types of drug toxicity were caused by genetic disorders affecting drug metabolizing enzymes and led Vogel (1959) 2 years later to coin the term "pharmacogenetics" to describe the interaction of heredity and environment in the pathogenesis of iatrogenic disease. Ecogenetics (Omenn and Motulsky, 1978) is concerned with hereditary factors that determine the pharmacological or toxicological responses to chemicals be these drugs, food constituents, or natural or manmade components of the environment. The difference between this discipline and ecogenetics is one of scale, as the same biochemical defense systems protect us from toxic injury by most chemicals, including drugs (Brieger, 1963; Brewer, 1971). Indeed, cytochrome P450 system and associated xenobiotic enzymes must be a crucial defense system against our chemical environment. Responses to drugs are the tip of the iceberg and are picked up because the responses usually are quick and the exposure is obvious (Calabrese, 1984, 1986). When differences in individual exposure are unknown and when the biological response is delayed, which it is bound to be in the case of neuronal cell death, links between individual susceptibility to certain chemicals and disease will be much harder to ascertain. In the 1960s a series of twin studies showed high heritability of the elimination of many different drugs (Vesell and Page, 1968a,bc; Vesell, 1982). However, it also became clear that it was a large multifactorial element with many environmental influences on the activity of these enzyme systems (Kalow, 1962,

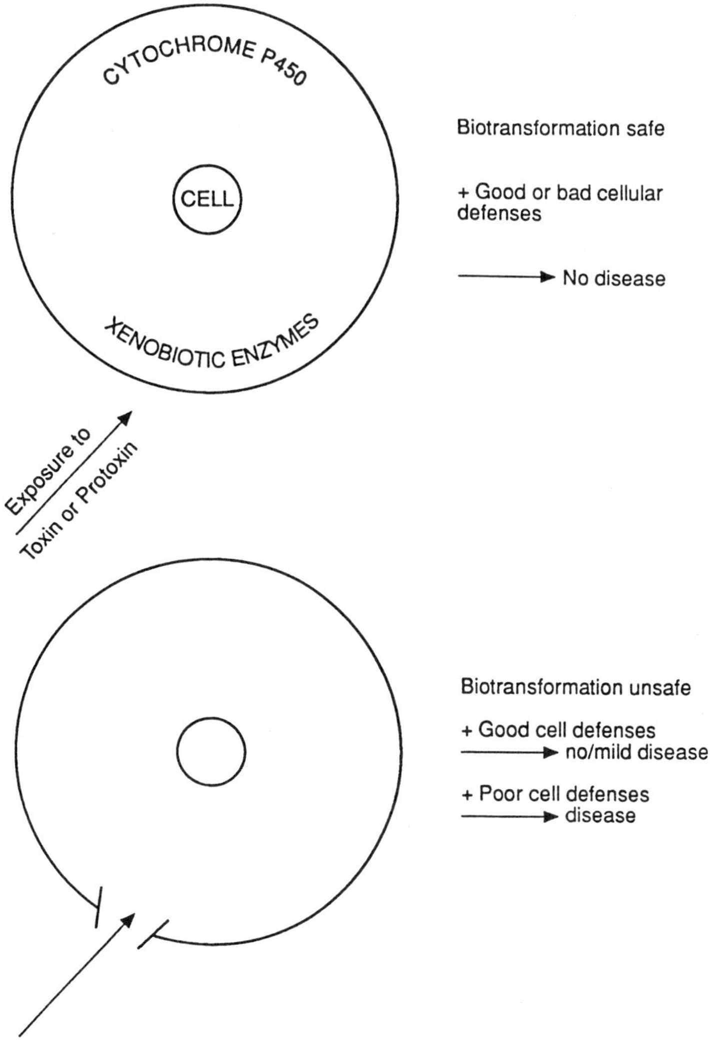

Figure 1 Interaction between safe and unsafe biotransformations and cellular defense systems. Currently topical defenses include uptake of chemical into cell or subcellular structures, ability to withstand oxygen and excitotoxic stress, including rises in intracellular calcium, and activation of proteolysis. The figure is diagrammatic—not anatomic—as important biotransformations can occur within neurons.

1965). These include the effect on the enzymes by the chemical itself best known from their inducibility or repression, both of which have clear teleologically interesting adaptive capabilities. Many of these enzyme systems are polymorphic, and a major part of the explanation for differences in activity within a given population and why incidence can vary extensively between different races, strains, and species (Kalow et al., 1986). Such variation must lead to a fitter population to cope with a changing chemical environment, although the evolutionary drive must have been greater to protect individuals up to the age of reproductive maturity and may not be as efficient for long-haul protection of neurons in senescence. The polymorphisms relevant to neurodegenerative disease may have survived from mutations that were neutral or advantageous in an evolutionary sense yet could be dangerous beyond the age of strict biological necessity.

Although we began our studies after a case of amyotrophic lateral sclerosis (ALS) occurring in close proximity to high exposure to the insecticide Permethrin (Pall et al., 1987; Steventon et al., 1990), a neurotoxin that kills insects by paralysis but not mammals who are protected by biotransformation to a nontoxic derivative, the same argument can be made from the dopaminergic neurotoxin, 1-methyl-4-phenyl-1,2,3,6-tetrahydropyridine (MPTP).

MPTP Revisited

MPTP poisoning remains a striking model of parkinsonism (Davis et al., 1979; Langston et al., 1983) and given that, in the laboratory acute experiments are done in the nonhuman primate is remarkably similar to idiopathic Parkinsonism. MPTP is a protoxin needing activation by monoamine oxidase B (MAO-B) to the toxin *N*-methyl-4-phenyl tetrahydropyridinium ion (MPP^+) (Chiba et al., 1984). If this reaction occurs outside the brain it is a detoxification reaction as MPP^+ cannot cross the blood–brain barrier. Most experiments in the laboratory and by drug addicts have bypassed first-pass metabolism by giving MPTP parenterally. However, it has been demonstrated that after oral administration extensive biotransformation occurs largely in the liver, such that very little MPTP reaches the brain and hence MPTP is essentially nontoxic by the oral route (Fuller and Hemrick-Luecke, 1987). Metabolism in liver is more complicated than brain, in that although the MAO-B pathway is active there is an alternative pathway that may even be preferentially used at low dosage using a variety of enzymes including cytochrome P450 to produce the pyridones of MPTP that are likely to be nontoxic and more water soluble and easily excreted (Wu et al., 1988). In the human epidemic of MPTP, individual susceptibility was not obvious but the epidemic was a small one and the dosage of MPTP administered often indeterminable. It is well established that there are marked differences between species susceptibility to MPTP and a variety of mechanisms postulated, perhaps relating mainly to amount of neuromelanin. Equally important, marked strain differences in mice have been described (Sonsalla and Heikkila, 1986, 1988). From these studies it was clear that much of the strain difference may lie in the degree of biotransformation of MPTP. In the resistant strains the amount of MPTP and consequently MPP^+ within brain was considerably less than in the susceptible strain. This did not appear to be due to differences in MAO-B activity and thus the alternative pathway identified in liver could be important. These experiments were done by giving MPTP subcutaneously, which would have largely avoided first-pass metabolism and the differences seen might be even larger if brain levels were studied after oral administration. Studies of these interstrain differences in the mouse or other species, given that there is some 80% homology between mouse and man, may give us crucial information and clues as to the basis of interindividual differences in toxicity of not only MPTP but many other xenobiotics with neurotoxic potential in the human and should be encouraged (Festing and Eydman, 1987).

There is no known source of MPTP but that does not mean that it does not exist and certainly many pyridines are present in diet (Maga, 1981). Importantly it can be formed from simpler pyridines that are known to exist by an N-methylation step (Ansher et al., 1986) using N-methyl transferases. It may well be important that nearly all the important MPTP analogues that are neurotoxins have a charged N-methyl group. Additionally, the more toxic tetrahydroisoquinolines and betacarbolines usually also have a charged N-methyl group (Drucker et al., 1990). N-methylation is a reaction catalyzed by widespread xenobiotic enzyme that has been identified in many tissues including brain and has wide substrate specificity like most xenobiotic enzymes.

Methylation

Methylation is a conjugation reaction first requiring the synthesis of S-adenosylmethionine and then transfer of the methyl group from this coenzyme to the acceptor molecule catalyzed by transmethylases. A number of endogenous chemicals are methylated, for example N-methylation of amines such as norepinephrine, serotonin, and histamine, the *O*-methylation of catecholamines and hydroxyindoles, and the *S*-methylation of thiol compounds (Weinshilboum, 1984, 1987; Weinshilboum et al., 1978; Weinshilboum and Sladek, 1980; Weinshilboum and Raymond, 1977). Methyl conjugation is a significant pathway in the biotransformation of many xenobiotics, although rarely emphasized compared with other reactions. However, it has long been known that methylation of pyridine to *N*-methylpyridinium hydroxide and selenate to dimethylselenide occurs. Methylation also occurs for arsenic tellurium and mercury (de Bruin, 1976). Several methyl-conjugating enzymes have been reported to have human polymorphisms. These enzymes are catechol-*O*-methyltransferase (COMT) (Raymond and Weinshilboum, 1975), thiopurine methyltransferse (TPMT), histamine *N*-methyltransferase (HNMT), and thiolmethyl-transferase (TMT) (Lennard et al., 1986). For all these enzymes high heritability has been demonstrated (Keith et al., 1983) as has a surprising degree of correlation between red blood cell enzyme activity and activity in organs such as the liver and kidneys. For COMT a significant correlation exists between red blood cells, COMT activity, and the proportion of *L*-dopa converted to 3-*O*-methyldopa (Reilly et al., 1980). TPMT catalyzes the *S*-methylation of thiopurine drugs such as mercaptopurine and azothioprine and have been correlated with production of toxic metabolites and toxicity of cephalosporine (Lennard et al., 1987; Kerremans et al., 1985).

Testing of Xenobiotic Enzymes

One method used is to test enzyme activity in readily available tissue, such as red blood cells. The degree of correlation appears good under some circumstances (Sladek-Chelgren and Weinshilboum, 1981) but has the disadvantages of testing only in available tissue and that it is an *in vitro* test. Another method that has been used widely is to give a compound by the oral route that is safe, ethically acceptable, and is thought to test a given pathway. This methodology has the advantage of being an *in vivo* test and a test of function of the enzymes in their natural supramolecular state. It is not yet possible, with the exception of the poor debrisoquine metabolizer, to test patients at a molecular level. Undoubtedly this will have advantages but may be complicated by multiple aberrations causing the same metabolic phenotypes so it is likely for the foreseeable future that even when such molecular probes become available they will need to be combined with more functional tests of enzyme activity. A combinal approach will also determine which polymorphisms at a DNA level have no functional consequences. Because of the known or unknown effects of other drugs on these pathways, one should be skeptical about any studies done on patients taking drugs or who have reached the stage of their disease where their diet may have altered significantly.

N-Methylation

Our group has studied this reaction *in vivo*, capitalizing on the well accepted observation that nicotinamide is N-methylated by an N-methylase that is known to be active on many substrates. The enzyme is thought to be the same as that which converts 4-phenylpyridine to MPTP. A pilot study suggests that patients with Parkinson's disease but not controls, or patients with ALS produce in urine after an oral dose very large amounts of N-methyl nicotinamide (Green et al., 1991; Table 1). Rather than an increase in N-methylation, further catabolism of N-methyl nicotinamide is probably blocked as its pyridone derivative was in very low concentration in the patients with Parkinson's disease. These pyridones correspond to the MPTP pyridones mentioned in the alternative pathway of biotransformation for MPTP present in the liver but not in the brain. It therefore seems possible that patients with Parkinson's disease have a tendency to develop a buildup of N-methylpyridines when challenged with a pyridine such as nicotinamide and by analogy probably a number of other pyridines and perhaps nonpyridines such as betacarbolines and tetrahydroisoquinolines. Given that these enzyme systems are thought to have high heritability, this suggests that this defect has antedated the onset of clinical disease and may be a significant risk factor. Nicotine is a pyridine and it is possible that the supposed protective effect of nicotine for the development of Parkinson's disease is because N-methyl nicotine happens to be nontoxic, and on a competitive basis reduces methylation of a more critical compound. It is also of note that nicotinamide in the form of NADH is a critical component of complex I and buildup of a closely related pyridine could inhibit this enzyme complex, and that nicotinic receptors are lost early in Parkinson's disease.

Table 1. Nicotinamide Metabolism in Parkinson's Disease Compared with Controls and Amyotrophic Lateral Sclerosis.

Group	Age (yrs)	No. of Patients	Nicotinamide (mg)	N′-Methyl-nicotinamide (mg)	N′-Methyl 2-pyridone-5-carboxyamide (mg)
Controls	61.8+18.0	31	10.9 (0.49–154.6)	2.4 (1.0–410.7)	18.3 (0.4–64.7)
Parkinson's disease	62.4+11.6	25	14.8 (1.0–391.2)	223.7[a]	7.7[b]
				(5.2–1001.0)	(0.2–38.9)
ALS	64.2+12.3	23	7.3 (1.1–495.8)	19.2 (1.0–1183.0)	20.3 (10.5–107.6)

ALS: Amyotrophic lateral sclerosis

Ages are mean + S.D.

Results are median values, with ranges in brackets, of mass of metabolite excreted in an 8-hr urine collection after 50 mg of nicotinamide given orally.

[a] $P < .05$ (Mann-Whitney test).

[b] $P < .005$ (Mann-Whitney test).

N-Methylation is catalyzed by an enzyme that uses many substrates other than nicotinamide. Some *N*-methylated pyridines, tetrahydroisoquinolines, and β-carbolines are more MPP^+-like than their parent compounds. This provides a mechanism whereby there may be individual susceptibility to such exogenous or endogenous compounds.

S-Methylation

We have studied the activity of TMT in red blood cell membranes of ALS and found high levels, whereas patients with Parkinson's disease tended to have low levels (Waring et al., 1989). In both groups of patients poor *S*-oxidation (Steventon et al., 1988, 1989) with raised cysteine sulfate ratios also were found (Heafield et al., 1990). These are both indirect methods of measuring the function of cysteine oxidase, which appears to be functioning poorly in a number of neurodegenerative diseases. The tendency for increased *S*-methylation, however, appears more specific for ALS. TMT is involved in the methylation of certain compounds, such as mercury, whose methyl derivative is more toxic than the parent compound. It is therefore possible that this pathway, like the *N*-methyl pathway in Parkinson's disease, leads to hypersusceptibility to some chemicals in patients destined to develop ALS.

Implications for Therapy

If true, there are clearly possibilities of reducing N-methylation in Parkinson's disease and S-methylation in ALS. Given the evidence of high heritability for these enzyme systems, it is also possible that picking up these defects will be possible using current or future methodology and treatment started at the at-risk phase. Treatment could clearly include reduction of environmental load of a critical compound or modification of enzyme activity toward greater safety, or competitively blocking the system by introducing a safe substrate. There also may be considerable implications for slowing disease progress in patients already diagnosed. This presupposes that by the time disease is evident that secondary processes are not already causing more damage than the primary defect if this is truly a multistage process.

Implications for Epidemiological Studies

Epidemiology so far has been rather unsatisfactory in that it has not come up with clear risk factors other than age for these conditions, although some weak but interesting associations have been described. However, if one could combine an epidemiological survey with a biochemical marker of hypersusceptibility there would be a striking increase in the power of the study. First of all, knowing the hypersusceptibility marker would give important clues as to what environmental triggers should be assessed. Second, and equally important, one could take a disease population and divide them biochemically into unsafe and safe metabolizers based on the relevant enzyme activity. The safe metabolizers would need, in order to develop disease, to have had a larger environmental burden of the relevant toxin or protoxin. By looking at that group and comparing them with controls and unsafe metabolizers, a correlation with exposure to a lifestyle or compound may become evident that would be hidden if one looked at the whole disease group. An alternative strategy would be to match a non disease group biochemically and compare with the disease group and then with both having the same xenobiotic enzyme profiles the supposition would be that the environmental stimulus would stand out in the diseased individuals.

References

Ansher SS, Cadet JL, Jakoby WB, Baker JK (1986): Role of *N*-methyltransferases in the neurotoxicity associated with the metabolites of 1-methyl-4-phenyl-1,2,3,6-tetrahydro-pyridine (MPTP) and other 4-substituted pyridines in the environment. *Biochem Pharmacol* 35:3359–3363.

Brewer GJ (1971): Annotations: human ecology, an expanding role for the human geneticist. *Am J Hum Genet* 23:92–94.

Brieger H (1963): Genetic bases of susceptibility and resistance to toxic agents. *J Occup Med* 5(11):511–515.

Calabrese EJ (1984): *Ecogenetics: Genetic Variation in Susceptibility to Environmental Agents.* New York: Interscience.

Calabrese EJ (1986): Ecogenetics: historical foundation and current status. *J Occup Med* 28(10):1096–1102.

Chiba K, Trevor A, Castagnoli Jr N (1984): Metabolism of the neurotoxic tertiary amine, MPTP, by brain monoamine oxidase. *Biochem Biophys Res Commun* 120:574–578.

Davis GC, Williams AC, Markey SP, et al. (1979): Chronic Parkinsonism secondary to intravenous injection of Meperidine analogues. *Psychiatr Res* 1:249–254.

de Bruin A (1976): *Biochemical Toxicology.* Amsterdam: Elsevier.

Drucker G, Raikoff K, Neafsey EJ, Collins MA (1990): Dopamine uptake inhibitory capacities of B-carboline and 3,4-Dihydro-B-Carboline analogues of *N*-Methyl-4-phenyl-1,2,3,6-tetrahydropyridine oxidation products. *Brain Res* 509:125–133.

Festing MFW, Eydman T (1987): Genetic variation in the response of mice to xenobiotics in vitro. General methodology and response to some model compounds. *Toxic in Vitro* 1:115–125.

Fuller RW, Hemrick-Luecke SK (1987): Oral versus parental efficacy of 1-methyl-4-phenyl-1,2,3,6-tetrahydropyridine (MPTP): differential effects on depletion of heart norepinephrine and striatal dopamine in mice. *Biochem Pharmacol* 36:789–792.

Green S, Buttrum S, Molloy H, et al. (1991): *N*-methylation of pyridines in Parkinson's disease. *Lancet* 338:120–121.

Heafield MTE, Fearn S, Steventon GB, Waring RH, Williams AC, Sturman SG (1990): Plasma cysteine and sulphate levels in patients with motor neurone, Parkinson's and Alzheimer's disease. *Neurosci Lett* 110:216–220.

Kalow W (1962): *Pharmacogenetics: Heredity and the Response to Drugs.* Philadelphia: WB Saunders.

Kalow W (1965): Contribution of hereditary factors to the response to drugs. *Fed Proc* 24:1259.

Kalow W, Goedde HW, Agarwal DP, eds. (1986): *Ethnic Differences in Reactions to Drugs and Xenobiotics.* New York: Alan R Liss.

Keith RA, Van Loon J, Wussow LF, Weinshilboum RM (1983): Thiol methylation pharmacogenetics: heritability of human erythrocyte thiol methyltransferase activity. *Clin Pharmacol Ther* 34:521–528.

Kerremans AL, Lipshy JJ, Van Loon J, et al. (1985): Cephalosporin-induced hypoprothrombinemia: possible role for thiol methylation of 1-methyltetrazole-5-thiol and 2-methyl-1,3,4-thiadiazole-5-thiol. *J Pharmacol Exp Ther* 235:

382–388.

Langston JW, Ballard P, Tetrud JW, Irwin I (1983): Chronic Parkinsonism in humans due to a product of meperidine-analog synthesis. *Science* 219:979–980.

Lennard L, Van Loon J, Lilleyman J, et al. (1986): Thiopurine pharmacogenetics: correlation of thiopurine methyltransferase (TPMT) activity and 6-thioguanine nucleotide (6-TGN) levels. *Clin Pharmacol Ther* 39:206.

Lennard L, Van Loon J, Lilleyman J, Weinshilboum RM (1987): Thiopurine pharmacogenetics in leukemia: correlation of erythrocyte thiopurine methyltransferase activity and 6-thioguanine nucleotide concentrations. *Clin Pharmacol Ther* 41:18–25.

Maga JA (1981): Pyridines in food. *J Agric Food Chem* 5:895–898.

Motulsky AG (1957): Drug reactions, enzymes, and biochemical genetics. *JAMA* 165:835–837.

Mountain JT (1963): Detecting hypersusceptibility to toxic substances. *Arch Environ Health* 6:357.

Niwa T, Takeda N, Kaneda N, Hashizume Y, Nagatsu T (1987): Presence of tetrahydroisoquinoline and 2-methyl-tetrahydro-isoquinoline in Parkinsonian and normal human brain. *Biochem Biophys Res Commun* 144:1084–1089.

Niwa T, Takeda N, Yoshizumi H, et al. (1991): Presence of 2-methyl-4-7-dihydroxy-1,2,3,4-tetrahydroisoquinoline and 1,2-dimethyl-4-7-dihydroxyl-1,2,3,4-tetrahydroisoqinoline, novel endogenous amines, in Parkinsonian and normal human brains. *Biochem Biophys Res Comm* 177:603–609.

Niwa T, Yoshizumi H, Tatematsu A, Matsuura S, Nagatsu T (1989): Presence of tetrahydroisoquinoline, a Parkinsonism related compound in foods. *J Chromatogr* 493:347–352.

Omenn GS, Motulsky AG (1978): Ecogenetics: genetic variation in susceptibility to environmental agents. In: *Genetic Issues in Public Health and Medicine*, Cohen BH, Likerfeld AM, Huany PC, eds. Springfield, IL: Charles C Thomas.

Pall HS, Williams AC, Waring RH, Elias E (1987): Motor neurone disease as manifestation of pesticide toxicity. *Lancet* ii:685.

Raymond FA, Weinshilboum RM (1975): Microassay of human erythrocyte catechol-O-methyltransferase: removal of inhibitory calcium with chelating resin. *Clin Chim Acta* 58:185–194.

Reilly DK, Rivera-Calimlin L, Van Dyke D (1980): Catechol-O-methyltransferase activity: a determinant of levodopa response. *Clin Pharmacol Ther* 28:278–286.

Sladek-Chelgren S, Weinshilboum RM (1981): Catechol-O-methyl-transferase biochemical genetics: human lymphocyte enzyme. *Biochem Genet* 19:1037–1053.

Sonsalla PK, Heikkila RE (1986): The influence of dose and dosing on MPTP-induced dopaminergic neurotoxicity in mice. *Eur J Pharmacol* 129:339–345.

Sonsalla PK, Heikkila RE (1988): Neurotropic effects of 1-methyl-4-phenyl-1,2,3,6-tetrahydroxypyridine (MPTP) and methamphetamine in several strains of mice. *Prog Neurol Psychopharmacol Biol Psychiatry* 12:345–540.

Steventon G, Heafield MTE, Waring RH, Williams AC (1989): Xenobiotic metabolism in Parkinson's disease. *Neurology* 39:883–887.

Steventon G, Waring R, Williams AC (1990): Pesticide toxicity and motor neuron disease. *J Neurol Neurosurg Psychiatry* 53:621–625.

Steventon G, Williams AC, Waring RH, Pall HS, Adams D (1988): Xenobiotic metabolism in motoneuron disease. *Lancet* i:644–647.

Vesell ES (1982): On the significance of host factors that affect drug disposition. *Clin Pharmacol Ther* 31:1.

Vesell ES, Page JG (1968a): Genetic control of dicumarol levels in man. *J Clin Invest* 47:2657–2663.

Vesell ES, Page JG (1968b): Genetic control of drug levels in man: phenylbutazone. *Science* 159:1479–1480.

Vesell ES, Page JG (1986c): Genetic control of drug levels in man: antipyrine. *Science* 161:72–73.

Vogel F (1959): Moderne Probleme der Humangenetik. (Modern problems of human genetics.) *Ergeb Inn Med Kinderheilk* 12:52–125.

Waring RH, Steventon G, Heafield MTE, Sturman SG, Williams AC, Smith M (1989): S-methylation in Parkinson's disease and motor neurone disease. *Lancet* ii:356–357.

Weinshilboum RM (1984): Human pharmacogenetics of methyl conjugation. *Fed Proc* 43:2303–2307.

Weinshilboum RM (1987): Pharmacogenetics of methyl conjugation and thiopurine drug toxicity. *Bioassays*, 7(2):78–82.

Weinshilboum RM, Raymond FA (1977): Inheritance of low erythrocyte catechol-O-methyltransferase activity in man. *Am J Hum Genet* 29:125–135.

Weinshilboum RM, Raymond FA, Pazmino PA (1978): Human erythrocyte thiopurine methyltransferase: radiochemical microassay and biochemical properties. *Clin Chim Acta* 85:323–333.

Weinshilboum RM, Sladek SL (1980): Mercaptopurine pharmacogenetics: monogenic inheritance of erythrocyte thiopurine methyltransferase activity. *Am J Hum Genet* 32:651–662.

Wu E, Shinka T, Caldera-Munoz P, Yoshizumi H, Trevor A, Castagnoli N Jr (1988): Metabolic studies on the nigrostriatal toxin MPTP in its MAO B generated dihydropyridinium metabolite $MPDP^{+}$. *Chem Res Toxicol* 1:186–194.

Chapter 15

The Natural History of Amyotrophic Lateral Sclerosis: Methodological Considerations

Marinos C. Dalakas

Amyotrophic lateral sclerosis (ALS) is a clinicopathologically distinct neuromuscular disease characterized by involvement of both the upper and the lower motor neurons. In its classic form it presents with diffuse fasciculating weakness and muscular atrophy, diffuse denervation, spasticity, hyperreflexia, and no signs of abnormal sensation, sphincteric dysfunction, or a systemic illness. Although in some patients the disease may be confined initially to the lower motor neurons (called progressive muscular atrophy, or progressive bulbar palsy), most patients develop subclinical or clinical signs of upper motor neuron involvement during the course of their illness (Williams and Windeback, 1991; Mulder, 1980). In this chapter, the term ALS will be conveniently used as an all-inclusive term covering all the clinical variants of what may be one single disease. This generalization, although appropriate in discussing pathogenesis, should be viewed with caution if strictly applied to the study of the epidemiology or natural history of the disease because clinical observations suggest that each clinical variant of ALS may have gender preference and different degree of progression or natural history, as discussed later.

ALS is a progressive disease that invariably leads to death after an average period of 3 to 5 years from onset. The natural history of the disease is essentially a reflection of the rate of deterioration of both the upper and the lower motor neurons. Although the exact rate of progression of ALS is critical for management, prognosis, or design of experimental

ADVANCES IN RESEARCH ON NEURODEGENERATION, I
D.B. Calne et al.

therapeutic drug trials, our main knowledge of the natural history of the disease had been based mostly on information derived from rather insensitive methods and scales that had produced subjective and ordinal data. It was not until 1988 when an objective and reproducible method of serial measurements of muscle strength was introduced based on a composite scoring system of the maximum voluntary isometric contraction (MVIC) of individual muscles (Munsat et al., 1988). This method, which has now been extended to include the bulbar muscles in a large number of patients (Tan et al., 1988; Brooks et al., 1991), is generating definitive data regarding the natural history of the disease. While this information is forthcoming, a number of methodological issues concerning the assessment of the natural history of the disease have emerged. I will discuss here the available information on the natural history of ALS, addressing at the same time the deficiencies in the methodologies employed.

Quantification and Natural History of the Upper Motor Neuron Loss

Current methods have addressed only the progression of the neurological deficit resulting from the loss of the lower motor neurons (i.e., muscle weakness and atrophy, fasciculations, muscle cramps and loss of amplitude of the muscle action potential recorded from an atrophying muscle upon electrical stimulation of its respective motor nerve). However, the component of the neurological deficit resulting from the loss of the upper motor neurons has been overlooked mainly because of lack of objective means to measure spasticity or pseudobulbar signs and our inability to distinguish which part of the muscle weakness in an affected extremity is directly related to the loss of the upper motor neurons. Further, it is not known whether the rate of progression of the lower motor neurons parallels that of the upper motor neurons, or the two progress independently. Sometimes, the development of spasticity in a weak, atrophic, and previously hypotonic limb seems to provide some support (acting like a cane), giving the false impression of temporary stability or even improvement. Other times, spasticity manifested early in the disease may be unnoticed by the patient because stiffness alone without atrophy, cramps, or definite muscle weakness could delay seeking medical attention.

In an effort to assess the *in vivo* status of the upper motor neurons in ALS patients, we evaluated the regional cerebral metabolic rate for glucose (rCMRGlu) by positron emission tomography (PET) with [^{18}F]2-

fluoro-2-deoxy-D-glucose (FDG) (Dalakas et al., 1987; Hatazawa et al., 1988). We performed 19 studies in 12 ALS patients and compared the rCMRGlu in 8 patients who had clinical signs of both upper and lower motor neuron disease (ALS/UMN) to 11 age-matched normals and to 4 patients who had progessive disease confined to the lower motor neurons (ALS/LMN). Three additional patients with old paralytic poliomyelitis who had severe residual lower motor neuron deficit (Dalakas, et al., 1986) also were studied for comparison.

We found that the mean rCMRGlu was lower by 21% in the ALS/UMN group compared with the age-matched control subjects $p < 0.1$ (Table 1). The degree of hypometabolism correlated with the duration of the clinical signs and extended throughout the cortex and the basal ganglia but not the cerebellum (Fig. 1). A two-sided Student's t test yielded a statistical significance of $p < .05$ for all regions except the cerebellum. Repeat studies done in four patients as their disease had progressed showed significant reduction in the metabolic rate for glucose corresponding to the worsening of the overall clinical picture (Fig. 2). When the cortical metabolic activity was correlated with the duration of the disease, a linear regression analysis yielded a 6% decline per year with $p < .05$ probability that the difference from zero slope was not due to chance (Fig. 3). In patients with ALS confined to the lower motor neurons (ALS/LMN) and in the patients with old paralytic poliomyelitis,

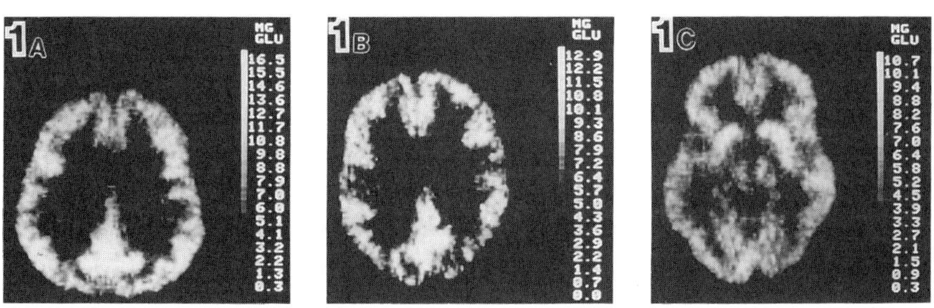

Figure 1 A: Scan of an ALS/lower motor neuron patient with progressive quadriparesis and bulbar involvement with disease confined to the lower motor neurons. The glucose utilization is within the normal range. B,C: Scans of an ALS/upper motor neuron patient with upper motor neuron disease shows marked metabolic depression diffuse in all cortical regions. (Color scales adjusted to the same metabolic range for quantitative comparison, show the striking difference in mg of glucose utilization per 100 g of tissue.)

Table 1. Regional Cerebral Metabolic Rates for Glucose[a]

Region	LMN	ALS/UMN
Cortex	$n = 7$	$n = 8$
Frontal	0.91 ± 0.15	0.80 ± 0.15[b]
Temporal	0.89 ± 0.16	0.78 ± 0.12[b]
Frontoparietal	0.87 ± 0.13	0.83 ± 0.17[b]
Parietal	0.90 ± 0.13	0.79 ± 0.12[b]
Occipital	0.87 ± 0.10	0.78 ± 0.13[b]
Average cortex	0.89 ± 0.13	0.79 ± 0.13[b]
Lower structures	$n = 6$	$n = 5$
Caudate	0.92 ± 0.15	0.77 ± 0.18[b]
Putamen	0.91 ± 0.19	0.78 ± 0.15[b]
Thalamus	1.02 ± 0.14	0.79 ± 0.21[b]
Pons	0.97 ± 0.17	0.85 ± 0.10[b]
Cerebellum	1.17 ± 0.22	1.02 ± 0.15[b]

[a] ± Standard deviation, divided by the appropriate average of normal volunteers. The normal group values are 1.0, by definition.

[b] Significance that value is statistically different from normal control subjects is $p < .05$, using the two-sided t-test. The LMN values are not significantly different from normal control subjects at the $p < .05$ level.

LMN, patients with only lower motor neuron signs (3 post polio, 4 ALS); ALS/UMN, amyotrophic lateral sclerosis patients with upper and lower motor neuron signs; n, number of patients.

the rCMRGlu was normal in the cerebellum but lower by 11% in the cortex and by 5% in the basal ganglia, not yielding significance ($p < .20$) in most cortical regions (except the frontal).

Because the hypometabolism in ALS/UMN is diffuse (as it is often seen in Alzheimer's disease) and extends to the basal ganglia (as in Parkinson's disease) without a concomitant atrophy in those regions, the neuropathological overlap among all three conditions may be more prominent than is generally realized. This strengthens the previously proposed hypothesis of a unified pathogenesis of ALS, Parkinson's disease, and Alzheimer's disease (Calne et al., 1986; Calne and Eisen, 1989; Appel, 1981).

To assess the progression of the UMN dysfunction and quantify the metabolic changes in the cortex, repeat studies were performed in

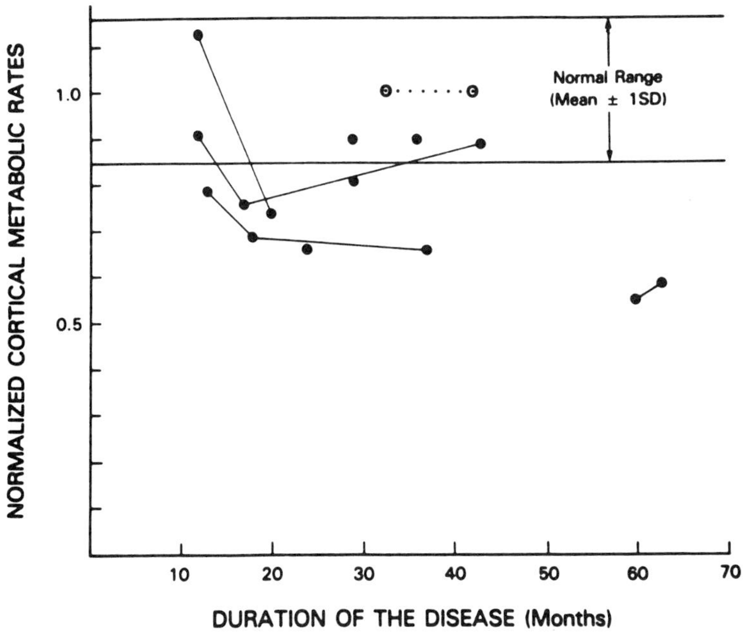

Figure 2 Average cortical metabolic rates for 14 ALS upper motor neuron studies (*solid circles*), normalized to the appropriate average of normal volunteers, versus duration of disease at time of the scan. The decline in metabolism with length of disease is significant at the $p = .05$ level. Four patients had sequential studies, as indicated by connecting lines. Contrasting with these sequential studies is a pair of sequential ALS/lower motor neuron studies (*open circles*), showing consistently normal metabolic rates despite rapid clinical worsening.

three patients with ALS/UMN and in one patient with ALS/LMN after a mean period of 11.2 months from the first PET scan (Dalakas et al., 1987). As shown in Figure 2, in the three ALS/UMN patients who were first studied 12 to 13 months after disease onset, the initial high rCMRGlu showed a striking decrease in repeat studies. In contrast, sequential studies on one ALS/LMN patient (open circles) showed normal and unchanged metabolic activity in the cortex despite a rapid clinical worsening of the LMN function that had led to quadriplegia.

Further analysis of the PET scans revealed that the decline of the upper motor neuron cortical metabolic activity did not follow the same pace from patient to patient. In one patient, for example, studied after 3 months from the first scan, no detectable changes were noted; in two other patients who had a normal first scan early in their disease, a 21% reduction was noted when restudied 12 months later; in another patient

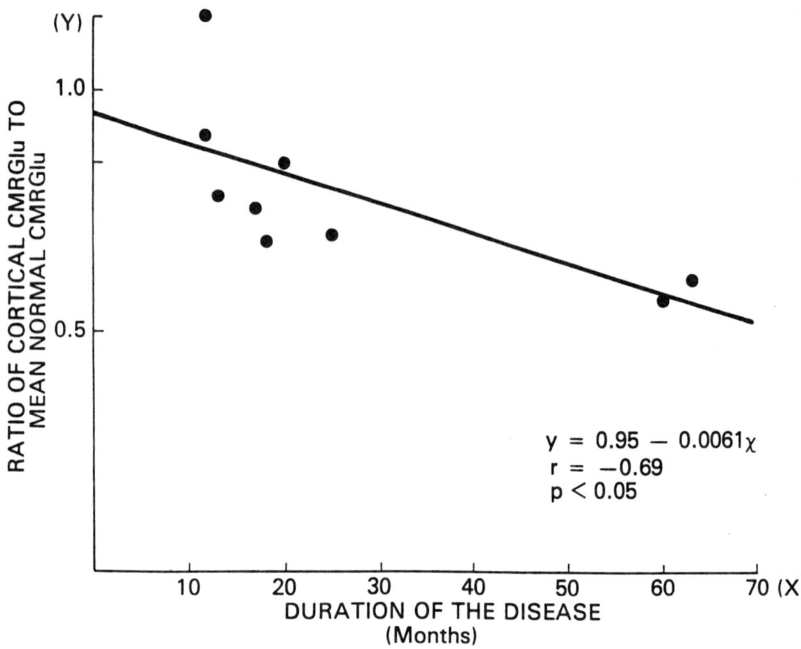

Figure 3 Correlation of the cortical metabolic activity of nine ALS patients with the duration of the disease. A linear regression analysis yields a 6% decline per year with a $p < .05$ probability that the difference from zero slope is due to chance.

restudied 2 years after the first scan only an additional 5% reduction of the rCMRGlu was noted; and yet in another patient, restudied after the same 2-year period, no further detectable reduction was noted despite severe clinical worsening (end stage of his disease). It appears, therefore, that in the same patient the deterioration of the UMN, as measured with PET scan, is linear over time and correlates with the duration of the disease but it does not keep pace with the rate of progression of the LMN. Instead, in certain stages of the disease the two progress independently. Future utilization of more sensitive scanning techniques and more specific ligants as tracers of neuronal metabolism will be helpful in providing accurate data separately for the progression of the UMN and LMN.

In agreement with clinical observations that ALS always begins asymmetrically (Brooks et al., 1991; Brooks, 1991), a significant right/left asymmetry also was observed in the cortical metabolic activity in several of our cases studied with the PET scan. Because the hypometabolism in one hemisphere did not always correspond to the site of the weakest or more atrophic extremity, multiple neuronal sites within both the UMN and the LMN pools may be concurrently affected in ALS patients.

The PET scan studies also have shown that in patients with ALS there is severe generalized hypometabolism even in the structurally normal cortex and in cortical regions traditionally thought to be unaffected, such as the frontal, parietal, or occipital cortex and the basal ganglia. This indicates that ALS is a more complex disorder than generally recognized and extends beyond the resolution of routine histopathology. The presence of hypometabolism in areas with histologically normal neurons suggests that an onging neuronal dysfunction of varying severity may occur long before the demise of these cells. Such cortical regions may contain neurons with normal perikaryon on routine histopathology but with diminished a number of dendritic spines and defective synaptic network when examined with special stains. Diminution or dissolution of a part of the elaborate dendritic arborization of the motor neurons has been demonstrated by Golgi stains in ALS spinal cord (Hirano, 1991) and can be of great functional significance in view of the numerous contacts made between the basilar dendrites with the other neighboring cells (Schiebel et al., 1974; Hammner et al., 1979). Each degenerated dendritic branch represents a loss of contact with at least four or five other large neurons via dendritic bundles (Schiebel et al., 1974). Because synapses in the brain also require consumption of glucose, the diminished number of dendrites and loss of synapses provide an explanation for the diffuse hypometabolism noted with the PET scans even in the anatomically "healthy" cortex.

The complex clinicoanatomical correlations and the widespread subclinical neuronal dysfunction unraveled by PET scan stress the difficulty in assessing the onset and natural history of the disease. Studying the dysfunction of the motor neurons *in vivo* while the cells are still alive is more informative than examining cell debris and gliosis, as done with the autopsied tissue. PET scan is a powerful tool that can assess quantitatively and serially the viability of motor neurons at rest, after passive movement of a weak extremity or even after electrical stimulation of a motor nerve, and provide useful data at the cellular level regarding the extent of the early lesion, the rate of progression, and the effectiveness of new therapies.

Progression of ALS According to Its Clinicoanatomical Variants

As discussed earlier, there are various forms of the motor neuron diseases group. When the disease is limited to dysfunction of the upper motor neurons it has very slow progression and the designation "primary lateral sclerosis" is preferred. It is still unclear whether primary lateral sclerosis is a distinct entity unrelated to the motor neuron diseases that will never become ALS, or if it represents an aborted form of ALS, or still if it is a slow form of ALS that has not yet clinically spread to the LMN.

When ALS is limited to the lower motor neurons of the brain stem nuclei, it is called "progressive bulbar palsy." Progressive bulbar palsy is more common in young females and, if such patients are treated aggressively to prevent aspirations, the spread of the disease to the upper or lower extremities can be slow (Munsat et al., 1988; Brooks et al., 1991). When the disease is limited to the lower motor neurons of the spinal cord it is designated as "progressive muscular atrophy" and it is characterized by slow progression. Careful follow-up examinations of such patients, however, reveals almost always some upper motor neuron signs. Observing, therefore, if the tendon reflexes are brisk, very brisk, present or hypactive is significant. In addition, some of these patients may have "conduction blocks" in proximal parts of their nerves or at sites of compression and represent a distinct entity "the multifaced motor neuropathy with conduction block." These patients may have antibodies to GM_1 goinglrosides and respond to immunotherapies. Last, when the disease presents with both upper and lower motor neuron signs, the designation of classic ALS is preferred. This usually has the fastest progression. The data available today has not addressed the natural history of the motor neuron loss separately for each clinical variant of ALS. Long-term follow-up of clinically homogeneous groups is needed to determine if these forms overlap and how often they do, or if they follow a different course from the beginning to end. This information is critical not only for better definition, prognosis, and management but also for the design of experimental therapeutic drug trials.

The Site of the Focal Onset may Influence Progression

The first clinical signs of ALS are remarkably asymmetric. Most of the patients appear to start with weakness of the right upper extremity and often with weakness of the distal hand (Brooks et al., 1991). The disease appears to spread faster within the contiguous areas at the original site of involvement. This local spread is faster at the brain stem, followed by the cervical region, and then by the lumbar region (Brooks et al., 1991). On the other hand, spread to noncontiguous areas is faster within the spinal cord than from the spinal cord to the bulbar region. Caudal-rostral symptomatic spread to a distant region usually depends on the distance of that region from the site of onset (Brooks et al., 1991). For example, the time for spread of the new weakness to the bulbar region is shorter after arm onset than leg onset. Overall, the leg strength deteriorates more slowly than the arm strength and a rostral-caudal spread is faster than the caudal-rostral involvement (Brooks et al., 1991). A sequential pattern of involvement is commonly that of leg-to-contralateral leg, or arm-to-contralateral arm.

Such a regional spread is of considerable interest because it resembles the pattern of spread of the paralysis in acute paralytic poliomyelitis where axonal transport has been implicated in the intraneuronal spread of the infection (Brooks, 1991; Jubelt et al., 1980). There is evidence that in acute poliomyelitis, once the poliovirus is inoculated naturally by a hematogenous spread or experimentally by direct injection into the central nervous system, the spread of the infection and the subsequent development of symptoms is due to axonal transport (Jubelt et al., 1980). In acute paralytic poliomyelitis, for example, the spread to the other leg appears to be significantly faster than the spread to either arm or to the bulbar region (Brooks et al., 1991). Because the pattern—but not the rate—of such a contiguous spread appears similar in both paralytic poliomyelitis and ALS, axonal transport also may play a role in the progression of symptoms (Gajdusek, 1985). There is a strong possibility that the poliomyelitis virus may enter the central nervous system (CNS) and reach motor neurons via retrograde axonal transport, perhaps originating at the motor endplate, which bears the poliovirus receptor (Leon-Monzon et al., 1992). If in ALS the disease spreads by means of axonal transport, the site of onset may provide clues regarding the type of neurotoxic elements that enter motor neurons via the most distal or exposed part of their axons. For example, when the disease starts at the bulbar region, neurotropic viruses or neurotoxic factors such as substances released from frequent muscle traumas may reach the regional motor neurons by entering through the distal motor axonal sprouts at the tongue, pharynx, or the mucosa of the facial muscles; when the disease begins at the hand, similar agents may enter through the distal motor axonal sprouts at the arm or hand. The neurotoxic potential of such hypothetical factors, the need for cumulative doses before they become pathogenic, and the type of axoplasmic flow involved can additionally influence the rate of progression and the aggressiveness of the disease. Because symptomatic weakness occurs when 70% to 80% of the motor neurons are lost in one spinal cord segment, the actual disease process probably begins years before the manifestation of the first symptoms, as discussed below.

Natural History as It Relates to Mortality

It is believed that 50% of the patients with ALS die within the first 3 years after onset of the first symptoms (Jablecki et al., 1989). However, the average length from the onset of the disease until death does not accurately represent the natural history of the disease. This is because the mortality in ALS patients is directly related to respiratory failure, which is in turn temporally related not only to the site of the initial disease onset (as discussed above) but also to other health-related problems such as cardiac or pulmonary. For example, the respiratory problems will begin sooner in an ALS patient who is also a chronic smoker and has emphysema, diabetes, or heart disease than in another ALS patient with an identical pattern of involvement but without similar health problems.

The Use of Maximum Voluntary Isometric Contraction in Assessing the Natural History of Motor Neuron Loss

All previous data on the natural history of ALS have been based on evaluating muscle strength with functional scales and manual muscle testing based on the MRC scale (Munsat et al., 1988). These methodologies are deficient, however, because the force in one extremity, especially when graded 4/5 in the MRC scale, may vary as much as 40%, and still fall into the same grade (Beasley, 1961). This insensitivity may account for the reported periods of stability in some patients with ALS (Mulder, 1980). Furthermore, the data obtained with the MRC scale are ordinal, and not suitable for parametric statistical evaluations (Munsat et al., 1988). Similarly, the functional scales are variable and insensitive because they are based on the difficulties with daily living, which vary from patient to patient according to their degree of adaptation, motivation, education, socioeconomic status, or associated depression. Recognition of these methodological problems has led to the use of MVIC as a tool of measuring strength utilizing electronic strain gauge tensiometers (Andres et al., 1988; Munsat et al., 1988). This method assures the objective quantification of muscle strength and, in conjunction with statistical analysis, offers a reliable and reproducible tool to assess the progression of the disease by measuring directly the output of the declining motor unit.

On the basis of the MVIC measurements, the course of ALS appears to be linear during the active phase of the illness, except for the very early and the very advanced stage (Munsat et al., 1988). Such a linear course is consistent with the process of continuous denervation, which occurs concurrently with that of reinnervation, as clearly shown on the muscle biopsies from the unaffected and affected extremities (Dalakas, 1989, 1990, 1991). In contrast, if the same number of motor neurons were lost per unit time without any degree of compensation in the form of reinnervation, the course of ALS would have been rapidly accelerated (Munsat et al., 1988). In the earliest phase of the disease, the motor neuron lose some of their distal axonal sprouts as a result of the sickness of the perikaryon (distal axonopathy). This is evidenced in the muscle biopsy of an unaffected (asymptomatic) extremity by the presence of small, scattered angulated fibers (Fig. 4a). However, the surrounding, still healthy, neurons compensate for the lost sprouts and rescue some of the newly denervated fibers resulting in enlargement of the motor unit, which is evidenced in the muscle biopsy by the presence of fiber-type grouping (Fig. 4b). Because the motor neurons in ALS have a short life span and limited reserves, they cannot support and maintain large motor units. This results in a continuous drop-out of muscle fibers, which is manifested in the muscle biopsy as "group atrophy" (Fig. 4c), and clinically corresponds to muscle weakness and atrophy (Dalakas 1977a,b).

---

a,b: Muscle biopsy from an *unaffected* extremity of a patient with ALS shows subclinical denervation in the form of small scattered angulated fibers (a) with some degree of reinnervation, as shown by formation of small groups of both fiber types (b); when the disease progresses further, the *weakening* extremity demonstrates dropout of whole motor units, seen in the muscle biopsy as group atrophy (c). In contrast, muscle biopsy from a *strong* muscle of a postpolio patient with fully effective, long-standing reinnervation depicts very large groups of both fiber types (d). Compare the degree of "effective" reinnervation between *strong* muscles in ALS (b) and postpolio (d).

See Figure 4 next page

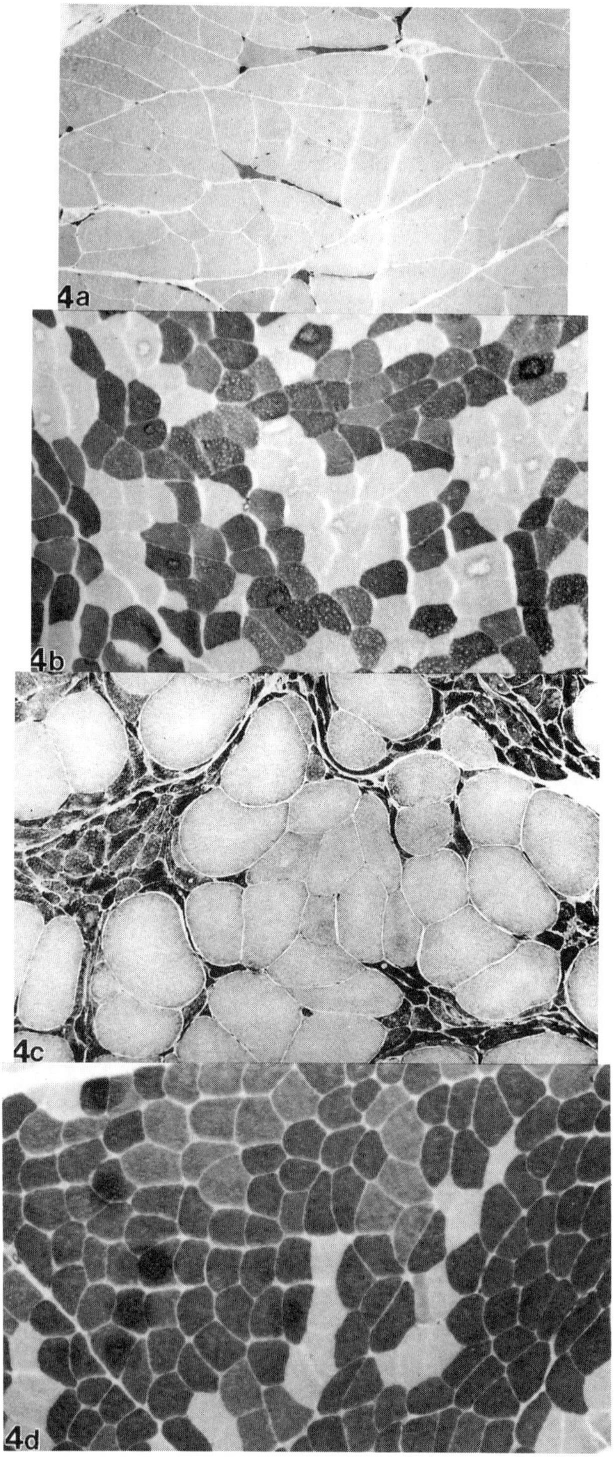
4a
4b
4c
4d

In ALS, the pace of muscle fiber drop-out and loss of motor units is different from the described in patients with the postpolio syndrome, who experience new muscle weakness 30 years after the original paralytic poliomyelitis (Dalakas et al., 1986). In postpolio patients, there is continuous denervation that keeps pace with a fully effective reinnervation from the surrounding neurons that are competent to oversprout and support large motor units (Dalakas and Hallett, 1988; Dalakas and Illa, 1991). This compensatory mechanism, which is seen in the muscle biopsy as very large fiber-type grouping (Fig. 4d), continues for approximately 25 to 30 years, when the pace of denervation begins to exceed a still effective reinnervation resulting in mild, slow, and erratically progressive muscle weakness (Dalakas et al., 1986). In contrast, in ALS the progression is rapid and linear because the ongoing steady-in-pace denervation far exceeds the pace of a not fully effective reinnervation.

Electrophysiological techniques appear to offer another tool to monitor the loss of the lower motor neurons. The amplitude of the muscle action potential recorded from the tibialis anterior muscle when the peroneal nerve is stimulated supramaximally in the popliteal fossa appears to show a linear correlation with the measurement of the MVIC (Kelly et al., 1990). When the changes in the amplitude of the muscle action potential are combined with the measurements of the MVIC, more precise information is obtained regarding the ongoing changes not only of the size, but also of the function, of the motor units in terms of the generated force. Whether the linearity of progression, as determined by MVIC alone, will be challenged by the electrophysiologic assessment of motor unit size when measured serially over time (Kelly et al., 1990) remains to be determined. Quantitative electromyography, when combined with single-fiber EMG, may prove to be a more powerful tool to assess disease progression directly at the motor unit level, by providing information not only on the loss of the lower motor neurons but also on the loss of their axonal sprouts and the effectiveness of reinnervation.

When Does ALS Start?

Because up to 70% or 80% of the motor neurons have to be lost before clinical weakness develops, asymptomatic denervation due to a smoldering dysfunction or loss of the lower motor neurons may be present for a number of years. As discussed above, the manifestation of weakness depends on the ratio of *denervation* (expressed as the number of disintegrating distal axonal sprouts or loss of whole motor neurons) to the compensated *reinnervation*, (expressed as the number of new axonal sprouts that make effective neuromuscular synapses). The fully reinnervated muscles are supplied by overtaxed motor neurons that can theoretically generate "normal" force for routine physical tasks, but may be insufficient to perform very demanding, repetitive, strenuous, or athletic physical activities. The reported higher incidence of ALS in some athletes (Kurtzke, 1991) may reflect a faster manifestation, or earlier recognition, of such an ongoing subclinical disease. An example supporting this hypothetical scenario is the subclinical illness of Lou Gehrig (Kasarskis and Winslow, 1989). About 1 1/2 years before he quit playing in May of 1939, Lou Gehrig's strength was normal and, although asymptomatic, he began the 1938 season at the lowest level of performance from any previous season (with −.3 cumulative deviation from the baseline batting average). This was remarkable for a player of his stature who had a consistently high batting average with cumulative deviation from the baseline +.343 (Kasarskis and Winslow, 1989). His hitting percentages also had declined and at the end of the 1938 season was down to −.8. His percentage of singles also demonstrated similar decline compared to previous years. However, his home run percentage did not decline until halfway through 1938 (Kasarskis and Winslow, 1989). During this period, Lou Gerhig had apparently normal strength and was able to make the movie "Rawhide." It is likely that during that year he had subclinical motor neuron disease that was not evident on everyday tasks but only in his athletic performance.

It can be hypothesized that chronic exposure of the motor neurons to a variety of viral, toxic, or degenerative processes may trigger a smoldering but well compensated motor neuron dysfunction. In such a subclinical condition, the muscles with normal strength may be supplied by the following different population of surviving motor neurons, as shown diagrammatically in Figure 5.

1. Normal, unaffected neurons, away from the area of lost neurons. These cells are not stressed to support reinnervation.
2. Normal, unaffected neurons, next to neighboring destroyed neurons. Such cells are stressed to oversprout and maintain larger than normal motor units in an effort to compensate for the lost neurons.
3. Neurons that were originally affected by the noxious agent(s) but have seemingly fully recovered. The life span of such cells may

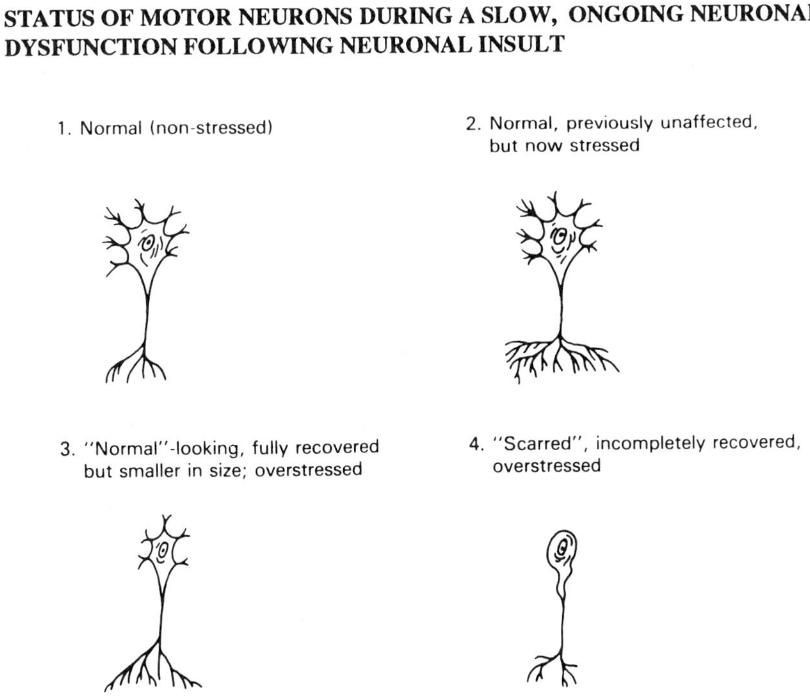

Figure 5 Status of various motor neurons during a smoldering and progressive, chronic motor neuron insult. Healthy neurons (1) coexist with stressed, reinnervating neurons that are either normal (2), "normal-looking" (3), or sick (scarred neurons) (4), which have limited reserves.

be less than normal and, if chronically stressed, their metabolic reserves may be limited.

4. Neurons that have been moderately affected but survived as smaller in size but normal in appearance. Such cells are dysfunctioning because, despite their normal appearance, they have impaired dendritic network and make less than normal number of effective synapses.
5. Neurons severely affected and incompletely recovered (scarred neurons) with very limited reserves and abnormal function.

In such a hypothetical situation the spinal cord contains less than the normal amount of normal neurons and an increased number of abnormal neurons that may not be able to perform to full capacity for a long period. As more cells are damaged by the original or new noxious agent(s), the pool of the remaining neurons declines further and when an up to 70%

to 80% of the neurons are lost, segmental weakness develops. A confounding factor in persons above the age of 60 years is the aging process (Tomlinson and Irving, 1977), which may accelerate the demise of the borderline-functioning neurons, leading to further depletion of the motor neuron pool and to earlier manifestation of weakness. This is consistent with the higher incidence of ALS in patients above the age of 60 years (Lilienfeld et al., 1989) and it is in accord with the hypothesis that motor neuron disorders, aging, and other neurodegenerative disorders may share a common basic process that leads to cell loss. It is relevant that signs of denervation with reinnervation (identical to those seen in ALS, as appear on Fig. 4a,b) occur in healthy muscles of normal volunteers above the age of 60 years (Dalakas, *unpublished observations*) as well as in clinically asymptomatic, healthy muscles of patients with Parkinson's disease and Alzheimer's disease (Dalakas, *unpublished observations*).

Can a Persistent Infection with a Lytic Virus Play a Role in an Apparent Neurogenerative Disorder such as ALS?

A classic example of motor neuron disease caused by a virus is the acute paralytic poliomyelitis in which motor neurons are infected with a lytic enterovirus of the picornaviridae family, the poliomyelitis virus. Although it has been traditionally thought that lytic viruses such as the poliomyelitis viruses cannot persist, recent evidence suggests that these viruses can be associated with a persistent infection if they escape immunological surveillance (Colbere-Carapin et al., 1989; Jubelt, 1991). This is accomplished if during the acute phase of the infection the virus-specific and MHC-restricted cytotoxic lymphocytes that are needed to clear the infection are ineffective or if an attenuated and not very lytic form of the virus manages to hide in cells that lack MHC expression and escape recognition (Oldstone, 1991). Such a classic cell is the neuron that lacks MHC expression and can serve as a safe house to harbor a virus. Neurons harboring such lytic viruses can stay intact for a number of years in the form of a safe symbiotic relationship, without generating inflammation or being subjected to phagocytosis and destruction. Furthermore, if the virion is not expressed on the surface of these cells, the antiviral antibodies may not detect it and cannot clear the infection.

The symbiotic relationship of the virus in these cells can change for a variety of reasons. The virus may mutate and become more potent, affecting the vital functions of the cell either rapidly or gradually. At first only the so-called luxury functions of the cell may be affected (Oldstone, 1991), such as the release of hormones or effective neurotransmitters or the ability of the motor cell to respond to stress by branching out to generate effective synapses. Gradually, the infection from cells that contain such mutant or defective viruses can spread via axoplasmic transport to neurons in contiguous segments of the spinal cord and the disease may begin spreading in a fashion analogous to the contiguous spread discussed earlier for the ALS and the acute polio. This hypothesis, which can explain the chronicity of the disease, the subclinical involvement, the suddenly rapid progression, and the lack of inflammation, may be applicable not only to ALS but also to Alzheimer's disease and Parkinson's disease. In Parkinson's disease there is already a natural model that develops several years after infection with a virus causing encephalitis lethargica (Von Economo, 1931). Patients with postencephalitic Parkinson's disease also have features of ALS with involvement of both the upper and the lower motor neurons (Hudson, 1991). It is possible that a similar virus may preferentially affect the basal ganglionic neurons responsible for dopamine secretion, as in Parkinson's disease, or the motor neurons responsible for voluntary motor movement, as in ALS. The overlapping features shared by both of these conditions, shown with the PET scan in ALS and with the muscle biopsies from the healthy muscles of Parkinson's disease, brings closer the suggested hypothesis of a unified mechanism (Calne and Eisen, 1989).

References

Andres PL, Finison LJ, Conlon T, Thisbodeau LM, Munsat TL (1988): Use of composite scores (megascores) to measure deficit in amyotrophic lateral sclerosis. *Neurology* 38:405–408.

Appel SH (1981): A unifying hypothesis for the cause of amyotrophic lateral sclerosis, parkinsonism and Alzheimers disease. *Ann Neurol* 10:499–505.

Beasley WC (1961): Quantitative muscle testing: principles and applications to research and clinical services. *Arch Phys Med Rehab* 42:398–425.

Brooks BR (1991): The role of axonal transport in neurodegenerative disease spread: a meta-analysis of experimental and clinical poliomyelitis compares with amyotrophic lateral sclerosis. *Can J Neurol Sci* 18:435–438.

Brooks BR, Sufit RL, DePaul R, Tan YD, et al. (1991): Design of clinical therapeutic trials in amyotrophic lateral sclerosis. In: *Advances in Neurology, Vol 56: Amyotrophic Lateral Sclerosis*, Rowland LP, ed. New York: Raven Press.

Calne DB, Eisen A (1989): The relationship between Alzheimer's disease, Parkinson's diseases and motor neuron disease. *Can J Neurol Sci* 16:547–550.

Calne DB, Eisen A, McGeer E, Spencer P (1986): Parkinson's disease, and motor neuron disease: abiotrophic interaction between aging and environment. *Lancet* ii:1064–1070.

Colbere-Carapin F, Christodoulou C, Crainic R, Pelletier I (1989): Persistent poliovirus infection of human neuroblastoma cells. *Proc Natl Acad Sci* 86.

Dalakas MC (1987a): ALS and post-polio: differences and similarities. In: *Research and Clinical Aspects of the Late Effects of Poliomyelitis, Vol 23*, Halstead LS, Wiechers DO, eds. White Plains: March of Dimes.

Dalakas MC (1987b): New neuromuscular symptoms after polio (the post-polio-syndrome): clinical studies and pathogenetic mechanisms. In: *Research and Clinical Aspects of the Late Effects of Poliomyelitis, Vol 23*, Halstead LS, Wiechers DO, eds. White Plains: March of Dimes.

Dalakas MC (1989): Post-polio motor neuron disease: what did we learn in reference to amyotrophic lateral sclerosis. In: *Motor Neuron Diseases*, Hudson A, ed. Toronto: University Press.

Dalakas MC (1990): Pathogenesis of post-polio syndrome: clues from muscle and CNS studies. In: *The Post-Polio Syndrome*, Munsat T, ed. Stoneham: Butterworths.

Dalakas MC (1991): The post-polio syndrome. *Curr Opinion Rheumatol* 2:901–907.

Dalakas MC, Hallett M (1988): The post-polio syndrome. In: *Advances in Contemporary Neurology*, Plum F, ed. Philadelphia: F.A. Davis.

Dalakas M, Illa I (1991): The post-polio syndrome: concepts in pathogenesis and etiologies. In: *Advances in Neurology, Vol 56: Amyotrophic Lateral Sclerosis*, Rowland LP, ed. New York: Raven Press.

Dalakas MC, Elder G, Hallett M, Ravitz J, Baker M, Papadopoulos N, Albrecht P, Sever JL (1986): A long term follow-up study of patients with postpoliomyelitis neuromuscular symptoms. *N Engl J Med* 314:959–963.

Dalakas MC, Hatazawa J, Di Chiro G, Brooks R (1987): Lowered cerebral utilization in patients with amyotrophic lateral sclerosis demonstrated by positron emission tomography (PET) scanning. *Ann Neurol* 22:580–586.

Gajdusek DC (1985): Hypothesis: interference with axonal transport of neurofilament as a common pathogenetic mechanism in certain disease of the central nervous system. *N Engl J Med* 312:714–719.

Hammner RP, Tomiyasu U, Scheibel AB (1979): Degeneration of the human Betz cell due to amyotrophic lateral sclerosis. *Exp Neurol* 63:333–346.

Hatazawa J, Brooks RA, Dalakas MC, Mansi L, Di Chiro G (1988): Cortical motor-sensory hypometabolism in amyotrophic lateral sclerosis: A PET study. *J Comput Assist Tomogr* 12:630–636.

Hirano A (1991): Cytopathology of amyotrophic lateral sclerosis. In: *Advances in Neurology, Vol 56: Amyotrophic Lateral Sclerosis*, Rowland LP, ed. New York: Raven Press.

Hudson AJ (1991): Amyotrophic lateral sclerosis/parkinsonism/dementia: clinico-pathological correlations relevant to Guamanian ALS/PD. *Can J Neurol Sci* 18:387–389.

Jablecki CK, Berry C, Leach J (1989): Survival prediction in amyotrophic lateral sclerosis. *Muscle Nerve* 12:833–841.

Jubelt B (1991): Viruses and motor neuron diseases. In: *Advances in Neurology, Vol 56: Amyotrophic Lateral Sclerosis*, Rowland LP, ed. New York: Raven Press.

Jubelt B, Narayan O, Johnson RT (1980): Pathogenesis of human poliovirus infection in mice. II. Age-dependency of paralysis. *J Neuropathol Exp Neurol* 39:149–159.

Kasarskis EJ, Winslow M (1989): When did Lou Gehrig's personal illness begin? *Neurology* 39:1243–1245.

Kelly JJ, Thibodeau L, Andres PL, Fenison LJ (1990): Use of electrophysiologic tests to measure disease progression in ALS therapeutic trials. *Muscle Nerve* 13:471–479.

Kurtzke JF (1991): Risk factors in amyotrophic lateral sclerosis. In: *Advances in Neurology, Vol 56: Amyotrophic Lateral Sclerosis*, Rowland LP, ed. New York: Raven Press.

Leon-Monzon M, Illa I, Dalakas MC (1992): Human muscle expresses poliovirus receptor (PVR) and can be infected with the poliovirus. *Neurology*, 42(S):229–230.

Lilienfeld DE, Chan E, Ehland J, Godbold J, Landrigan PJ, Marsh G, Perl DP (1989): Rising mortality from motor neuron disease in the USA, 1962–1984. *Lancet* i:710–713.

Mulder DW, ed. (1980): *The Diagnosis and Treatment of Amyotrophic Lateral Sclerosis.* Boston: Houghton Mifflin.

Munsat TL, Andres PL, Finison LT, Conlon T, Thibodeau L (1988): The natural history of motor neuron loss in amyotrophic lateral sclerosis. *Neurology* 38:409–413.

Oldstone MBA (1991): Molecular anatomy of viral persistence. *J Virol* 65:6381–6386.

Schiebel ME, Davies TL, Lindsay RD, Schiebel AB (1974): Basilar dendritic bundles of giant pyramidal cells. *Exp Neurol* 42:307–319.

Tan YD, Dolan P, Brooks BR (1988): Symptom progression in amyotrophic lateral sclerosis. *Ann Neurol* 24:14 (Abstract).

Tomlinson BF, Irving D (1977): The numbers of limb motor neurons in the human lumbosarcal cord throughout life. *J Neurol Sci* 34:213–219.

Von Economo C (1931): *Encephalitis Lethargica: Its Sequelae and Treatment.* London: Oxford University Press.

Williams DB, Windebank AJ (1991): Motor neuron diseases (amyotrophic lateral sclerosis). *Mayo Clin Proc* 66:54–82.

Part V

DISCUSSIONS

The papers presented in the first IWCN Symposium were followed by discussions. Four young neurologists were invited to record these exchanges or react to them. In some case, their text provides an account of what transpired; in others, they comprise a response to the stimulus generated by the discussions. The material generated by these four young neurologists are appended in the following pages as a brief epilogue.

Chapter 16

Normal Aging of the Nervous System

K. W. Lange and G. M. Paul

When studying age-related alterations in the brain it is important to examine the neurobiological parameters under investigation across the entire life span of humans and experimental animals rather than to confine investigations to limited stages of aging. Because of the scarcity of brain tissue from young humans, studies on human aging are normally confined to the adult age range, whereas animal studies usually do not include very old animals. The exclusion of aged animals may produce a false picture of aging. For example, numbers of neurons in cortical area 17 of the Sprague-Dawley rat do not alter until the animal reaches the age of 13 months (Peters et al., 1983). However, there may be neuronal loss in this area in 47-month-old diet-restricted rats (Peters et al., 1987).

Research on neurobiological changes over the whole life span would provide information on the development, in particular the maximum and minimum values, of cell numbers, receptor densities, etc., whereas comparisons between only two points of aging (e.g., early and late in life) may lead to erroneous conclusions. Various studies have shown that not all age changes progress steadily. A two-point study may give the false impression of a steady decline and may miss significant alterations occurring between youth and old age. In regard to cholinergic brain regions both a loss of neurons (Mann et al., 1984; McGeer et al., 1984) and no loss (Whitehouse et al., 1983; Chui et al., 1984) have been reported in the nucleus basalis of Meynert. An important difference between these results is that the studies reporting an age-related neuronal loss included data from children, whereas the studies that found no loss with age included only adults aged over 25 years. A decline associated with age in

ADVANCES IN RESEARCH ON NEURODEGENERATION, I
D.B. Calne et al.

the number of neurons in the human nucleus basalis may therefore occur before adulthood, or alternatively may take place throughout life but may be significant only when the period of life before adulthood is included in the analysis.

The comparison of two stages of aging is particularly problematic where the young subjects are still in the process of alterations of early development. Some early developmental changes of the rat hippocampus, for example, may not be complete until 12 months of age (Bayer, 1982).

The lack of neuronal loss in a given brain region does not necessarily signify the absence of age-related alterations. Since neuronal dendrites may represent up to 95% of the surface of a neuron that may come into contact with other neurons (Schade and Baxter, 1960), their growth and maintenance is important for the adaptations of the neuron to the loss of surrounding neurons and the integrity of the brain during aging. It is therefore essential not only to determine the number of neurons but also to assess the integrity of their dendritic arborization.

A further problem in the evaluation of age-associated changes in the central nervous system is the classification of neurons according to size and the differentiation between neurons and glial cells. It is possible that instances of loss of large neurons may reflect shrinkage of the cells, thus placing them in a different size category. For example, large and small neurons in the cortex have been examined in the course of aging using an automated image-analyzing system (Henderson et al., 1980). The greatest decrease in neuronal density was found in the category of large neurons. This result appears to be at variance with the greater density reduction of the predominantly small neurons in layers II and IV (Brody, 1955). This inconsistency could be the result of shrinkage rather than loss of large cells with aging since Brody's study (1955), in which cells were counted without regard to size, would not have found a shrinkage of large neurons. By contrast, the analysis of Henderson et al. (1980) could have been less sensitive to the loss of small neurons by including them in the glia category since the automatic classification into small neurons or glial cells was set by size. Once it is established whether the small cells counted are neurons or proliferating glia, the question of what it means for a neuron to shrink remains to be resolved. At present it is not clear whether shrunken neurons are functionally unaltered, impaired, or moribund.

A major problem when studying the effects of aging on the brain is to differentiate between age-related alterations and the influence of pathological changes associated with neuropsychiatric diseases. It is established that most persons over the age of 65, while appearing to be in good physical and mental health, show classical neuropathological alterations such as senile plaques and neurofibrillary tangles, which are usually associated with Alzheimer's disease, or Lewy bodies, which are seen in Parkinson's disease (Tomlinson et al., 1968; Dayan, 1970; Ball, 1977; Ulrich, 1985; Gibb, 1986; Mann et al., 1987). Most studies of age-related changes in the human brain have not excluded subjects with these histopathological alterations but rather have taken a general cross-section of the aged population. The typical neuropathological features of Alzheimer's disease and Parkinson's disease were regarded in these studies as nonexistent or minimal. It is to be expected therefore that the brain regions or neuronal cell types that are damaged in Alzheimer's disease or Parkinson's disease are those that appear to be susceptible to aging. These regions include the nucleus basalis of Meynert (Mann et al., 1984), the locus ceruleus (McGeer et al., 1977; Vijayashankar and Brody, 1979; Tomlinson et al., 1981), the substantia nigra (McGeer et al., 1977; Mann et al., 1984), the hippocampus (Ball, 1977; Miller et al., 1984; Mann et al., 1985), and the cerebral cortex (Brody, 1955; Henderson et al., 1980; Mann et al., 1985). By contrast, those brain areas such as the inferior olive, abducens, trochlear, and cochlear nuclei, which are unaffected in Alzheimer's disease and Parkinson's disease, are spared damage from aging (Konigsmark and Murphy, 1972; Monagle and Brody, 1974; Vijayashankar and Brody, 1977a,b). Cell loss that has been claimed to be associated with aging per se may be the result of mild cases or early stages of neurological or psychiatric diseases, which are increasingly present in the elderly. It may therefore be inappropriate to include subjects with senile plaques, neurofibrillary tangles, or Lewy bodies in studies on aging, since they may be affected by neuropsychiatric disease at a stage where clinical signs are not present.

The exact relationship between plaques and tangles and Alzheimer's disease and between Lewy bodies and Parkinson's disease is not yet clear. For example, the emphasis that has been placed on correlating the diagnosis of Parkinson's disease with the presence of Lewy bodies does not appear to be justified. Recent studies have shown that Lewy bodies occur in a small percentage of normal elderly individuals and to various extents in other neurodegenerative disorders, whereas not all patients with the clinical diagnosis of classical idiopathic parkinsonism appear to have Lewy bodies (Calne and Eisen, 1989). Similarly, senile plaques and neurofibrillary tangles may be absent in the neocortex of a substantial proportion of demented aged people, while a significant number of mentally intact elderly patients show these neuronal lesions and cannot in this respect be distinguished clearly from demented individuals (Jellinger, 1992). The question remains whether Alzheimer's disease and Parkinson's disease represent an exaggeration of normal aging or whether they are diseases caused by factors not normally active during the course of aging. Further research must be undertaken with a view to developing new classification systems for the diagnosis of Alzheimer's disease and Parkinson's disease based on etiology, pathology, clinical features, natural history, and response to therapy, and to elucidating the borders between normal aging and neurodegenerative diseases. In this context, it is interesting to question whether available antiparkinsonian drug therapies such as *L*-dopa, *L*-deprenyl, and dopamine agonists may influence the normal aging process.

References

Ball MJ (1977): Neuronal loss, neurofibrillary tangles and granulovacuolar degeneration in the hippocampus with ageing and dementia. A quantitative study. *Acta Neuropathol (Berlin)* 37:111–118.

Bayer SA (1982): Changes in the total number of dentate granule cells in juvenile and adult rats: a correlated volumetric and ^{3}H-thymidine autoradiographic study. *Exp Brain Res* 46:315–323.

Brody H (1955): Organization of the cerebral cortex. III. A study of aging in the human cerebral cortex. *J Comp Neurol* 102:511–556.

Calne DB, Eisen A (1989): The relationship between Alzheimer's disease, Parkinson's disease and motor neuron disease. *Can J Neurol Sci* 16:547–550.

Chui HC, Bondareff W, Zarow C, Slager U (1984): Stability of neuronal number in the human nucleus basalis of Meynert with age. *Neurobiol Aging* 5:83–88.

Dayan AD (1970): Quantitative histological studies in the aged human brain. I. Senile plaques and neurofibrillary tangles in normal patients. *Acta Neuropathol (Berlin)* 16:85–94.

Gibb WRG (1986): Idiopathic Parkinson's disease and the Lewy body disorders. *Neuropathol Appl Neurobiol* 12:223–233.

Henderson G, Tomlinson BE, Gibson PH (1980): Cell counts in human cerebral cortex in normal adults throughout life using an image analysing computer. *J Neurol Sci* 46:113–136.

Jellinger KA (1992): Morphology of the aging brain and relations to Alzheimer's disease (this volume).

Konigsmark BW, Murphy EA (1972): Volume of the ventral cochlear nucleus in man—its relationship to neuronal population and age. *J Neuropathol Exp Neurol* 31:304–316.

Mann DMA, Yates PO, Marcyniuk B (1984): Changes in nerve cells of the nucleus basalis of Meynert in Alzheimer's disease and their relationship to ageing and to the accumulation of lipofuscin pigment. *Mech Aging Dev* 25:189–204.

Mann DMA, Yates PO, Marcyniuk B (1985): Some morphometric observations on the cerebral cortex and hippocampus in presenile Alzheimer's disease, senile dementia of Alzheimer type and Down's syndrome in middle age. *J Neurol Sci* 69:139–159.

Mann DMA, Tucker CM, Yates PO (1987): The topographic distribution of senile plaques and neurofibrillary tangles in the brains of non-demented persons of different ages. *Neuropathol Appl Neurobiol* 13:123–139.

McGeer PL, McGeer EG, Suzuki JS (1977): Aging and extrapyramidal function. *Arch Neurol* 34:33–35.

McGeer PL, McGeer EG, Suzuki J, Dolman CE, Nagai T (1984): Aging, Alzheimer's disease, and the cholinergic system of the basal forebrain. *Neurology* 34:741–745.

Miller AKH, Alston RL, Mountjoy CQ, Corsellis JAN (1984): Automated differential cell counting on a sector of the normal human hippocampus: the influence of age. *Neuropathol Appl Neurobiol* 10:123–141.

Monagle RD, Brody H (1974): The effects of age upon the main nucleus of the inferior olive in the human. *J Comp Neurol* 155:61–66.

Peters A, Feldman ML, Vaughan DW (1983): The effect of aging on the neuronal population within area 17 of adult rat cerebral cortex. *Neurobiol Aging* 4:273–282.

Peters A, Harriman KM, West CD (1987): The effect of increased longevity, produced by dietary restriction, on the neuronal population of area 17 in rat cerebral cortex. *Neurobiol Aging* 8:7–20.

Schade JP, Baxter CF (1960): Changes during growth in the volume and surface area of cortical neurons in the rabbit. *Exp Neurol* 2:158–178.

Tomlinson BE, Blessed G, Roth M (1968): Observations on the brains of non-demented old people. *J Neurol Sci* 7:331–356.

Tomlinson BE, Irving D, Blessed G (1981): Cell loss in the locus coeruleus in senile dementia of Alzheimer type. *J Neurol Sci* 49:419–428.

Ulrich J (1985): Alzheimer changes in non demented people younger than 65: possible early changes of Alzheimer's disease and senile dementia of Alzheimer type. *Ann Neurol* 17:273–277.

Vijayashankar N, Brody H (1977a): Aging in the human brain stem. A study of the nucleus of the trochlear nerve. *Acta Anat (Basel)* 99:169–172.

Vijayashankar N, Brody H (1977b): A study of aging in the human abducens nucleus. *J Comp Neurol* 173:433–438.

Vijayashankar N, Brody H (1979): A quantitative study of the pigmented neurons in the nuclei locus coeruleus and subcoeruleus in man as related to aging. *J Neuropathol Exp Neurol* 38:490–497.

Whitehouse PJ, Parhad IM, Hedreen JC, Clark AW, White III CL, Struble RG, Price DL (1983): Integrity of the nucleus basalis of Meynert in normal aging. *Neurology* 33 (Suppl 2):159.

Chapter 17

Parkinson's Disease

Fabrizio Stocchi

Poewe: Donald, would you think that drug response would be important in the definition of PD in the presence of one or more clinical signs?

Calne: Drug response is not an essential but an additional consideration. There are accumulating reports of patients who have responded, and who have pathologically other disorders that are clearly distinct from PD, such as PSP. These patients in the early stages may respond. Others who resemble PD in their late stages, but at autopsy have tangles rather than Lewy bodies in the substantia nigra. Of course they raise an issue that is an important question. What do you do with a patient who looked exactly like PD during life, including response to treatment, but at post mortem you find dopamine loss in the brain and tangles. How is this patient classified?

Jellinger: It is my belief that PD is also a multisystem disorder, at least in the late stages. The so-called diagnostic markers (degeneration of substantia nigra and Lewy bodies) are not specific changes, and this is the same for the so-called Alzheimer pathology; we know from the English study group and ours that about 70% to 80% of all cases that clinically presented as idiopathic PD are confirmed by the current pathological criteria, the others are not. For example, two cases have been recently published who clinically looked like PD with mild dementia but at the pathology they had Alzheimer's disease changes, with a very severe damage of the substantia nigra. Other authors have recently shown that there is severe nigral neurons'loss in Alzheimer's disease ranging up to 50%. In conclusion, even at the post mortem it is extremely difficult to define idiopathic PD. We deal with a number of degenerative

ADVANCES IN RESEARCH ON NEURODEGENERATION, I
D.B. Calne et al.

disorders mainly involving the neustriatum dopaminergic system and we can distinguish some subgroups, e.g., the akinetic form from the tremor-dominant form, based on the involvement of the substantia nigra and the extent and regional distribution of cell loss within this area.

Finch: Let me address another factor in the definition of these diseases, i.e., the variable picture of neural loss in many regions as a function of aging. As far as I can tell, this is subject to different environmental factors and genetic influences and results in individual patterns of aging in quite a variable way from one individual to another, often interfering with the diagnosis of PD we are trying to establish. In fact, there is about 30% to 50% of neuronal loss in the substantia nigra at the age of 70 and this is true for other regions of the brain.

Olanow: Let me just briefly mention that in our NMR study we were able to identify patients in whom there was signal hypodensity in the putamen that we felt was the NMR equivalent of striatal degeneration. Interestingly, we saw this in about 20% to 30%. This subgroup clinically had a presentation with dominant involvement of speech and balance, rapid progression, and little or no response to levodopa. The other group that we felt to have nigral syndrome clinically had unilateral presentation, prominent resting tremor, and a good response to L-DOPA. So even if there is an overlap of parkinsonian syndromes, at least we have to take into account these two clinically different forms of the disease.

Calne: I think this is very helpful. Knowing what you are dealing with when you conduct epidemiological studies is of paramount importance. Part of the difficulty in epidemiological studies may well derive from the problem of separating subgroups. Prof. Jellinger, do you think it would be possible to formulate criteria for the possible, probable, and definite identification of PD under the microscope?

Jellinger: The current diagnostic criteria for, let us say, primary nigral degeneration or degeneration of the striatal-nigral, mesostriatal system is selective or very severe neuronal loss in the substantia nigra, pars compacta, particularly involving the ventrolateral and more the caudal and the rostral part of substantia nigra, with more or less intact striatum. Morphologically, it includes the occurrence of Lewy bodies mainly in the brain stem with a small number in the cortex and additional lesions of specific subcortical systems, particularly the serotonergic raphe system and the noradrenergic system of the locus coeruleus with more or less preservation of the cholinergic system. A small damage of the nucleus basalis of Meynert can be present. The efferent striatal systems are more or less intact and that explains the response to treatment. This is a major difference, for example, to the striatal-nigral degeneration where there is a complete lesion of the striatal-nigral system.

We believe that in the striatal-nigral degeneration, the damage of the substantia nigra is secondary to a primary damage of the striatum whereas in idiopathic PD, I believe, the primary damage is in the melanin-containing nigral neurons. Progessive supranuclear palsy [PSP] as a rule can be very well identified even if there are some cases with a lot of tangles that do not fulfill all the current criteria of PSP. These cases can be specific form of Alzheimer with severe subcortical involvement, a separate entity, or an atypical form of PSP. With respect to Dr. Finch's comment, I think that we overestimate the old studies reporting cell loss in normal aging. In fact, in recent studies, we did not find age-related cell loss in the nucleus basalis.

Finch: What I had tried to communicate is that there is the possibility of a statistical error when we look at the whole population, but this is not applicable to the individual case.

Jellinger: There is a threshold of cell loss in pathology. In fact, in PD there is a 70% to 80% dopaminergic neuron loss, and this is true for Alzheimer's disease with the cholinergic system. We found a 60% to 80% cell loss in the cholinergic system of demented PD too.

Olanow: Do you think that it is possible that you start with one event in a condition like PD, for example, loss of neurons in the substantia nigra, and subsequently to that loss a cascade of events occurs, for example, a disinhibition of a glutamatergic system, which leads to a glutamate overactivity resulting in toxicity to the striatum. What one sees ultimately is an evolving sequence of events not necessarily caused by the same original factor but by neurodegeneration occurring as a consequence of the initial disease process. This may help to understand why all these conditions, ultimately, are so similar, despite different etiology.

Snow: We had an experience with one patient when we were collecting our normal subjects. He had very low fluorodopa PET scan uptake. At that time he was 52, and about 18 months later he developed sufficient signs for us to be confident that he was developing PD. So, regarding the question of how long before clinical symptoms the diagnosis can be made by PET: for this patient, 9 to 12 months.

Calne: I would like to add a comment about this patient: at the time of diagnosis he had signs but no symptoms. Indeed, he still maintains quite firmly that there is nothing wrong with him.

Youdim: I would like to mention the possible relation between stress and PD. A colleague of mine observed a high incidence of PD within the Air Force pilots who participated in the war.

Poewe: I found stress can unmask or trigger the symptoms of PD. There are some reports about soldiers who developed a syndrome very similar to PD and there was another report that the incidence of PD within the war prisoners was much higher than in the normal population. Moreover, we all know that most patients report a stressful situation as a trigger for the development of Parkinson's disease.

Horowski: Is it possible to unmask symptoms by giving pharmacological compounds such as tetrabenazine, which depletes central dopamine?

Poewe: Rajput published two cases who developed a parkinsonian syndrome during neuroleptic treatment and showed at the post mortem examination Lewy-body–positive nigral degeneration.

LeWitt: There is an animal model of a stress-related dopaminergic deficiency syndrome that is well known to American Midwest farmers, where under stress of going to the market pigs can develop excessive salivation with a rigid state and episodic freezing. Studies have shown that there seems to be a decreased concentration of striatal dopamine relative to normal. I think these studies need further investigation because the animal returns to normal and this may be due to a sort of deficiency in dopamine synthesis. There is at least one case in the literature that carbachol infusion into the brain for treatment of Alzheimer's disease unmasks parkinsonism. Thus, it would be sensible to test, e.g., physostigmine in a large population in an attempt to unmask early symptoms of PD.

Mizuno: Regarding the usefulness of personality traits for a preclinical diagnosis of PD, I believe there are a lot of people with personality deviation who never develop PD. Thus, I believe we need a biochemical marker for preclinical diagnosis. Many years ago it was shown, in Japan, that L-DOPA absorption in the gastrointestinal tract is different between normal subjects and parkinsonian patients.

Youdim: Sandler, Stern, and I did a study and there was a significant difference in sulfate conjugation of dopamine between parkinsonian and normal subjects. But then, we ran into the problem that the same difference was found in depressed patients. So we cannot say that it is specific for PD.

Dalakas: I am wondering whether the preclinical phase of PD consists of a certain degree of compensation for the loss of striatal nigral neurons. The analogy could be with motoneuron disease, where we know that there is a loss of neurons but there is a certain degree of compensation provided by surviving neurons. My question is: Is there any sprouting from the surviving neurons?

Jellinger: I am not aware of any study demonstrating such sprouting; this is probably due to the loss of tyrosine hydroxylase even in neurons, which seem to be normal. You can have then a normal number of cells but they may not function properly.

Leigh: I was very interested to look after one patient in David Brook's study on the monozygotic twins. One of the twins presented with classical parkinsonism with asymmetrical resting tremor, which one would predict to have Lewy bodies pathology. The other twin presented about 5 years later with orthostatic tremor, which I think most of us would not recognize as typical of PD. And I think, coming back to the definition of "pure population," this is very important for pharmacological studies or drug trials, but for epidemiological studies we need broader definitions because one does not want to ignore patients who have the same molecular pathogenesis. We then need to define pharmacological criteria or diagnostic criteria for clinical trials, and perhaps other criteria for epidemiological studies on molecular genetics.

Olanow: To be practical I propose that considering the high percentage of patients with early PD who have smell deficiency, the sniff test should be done as well as the reaction test, which is abnormal in patients with early PD. These are very easy tests to do that may be validated in the future.

LeWitt: We studied tyrosine metabolism in PD and controls to see if the stimulation of biopterin synthesis could be a marker and we could not find any difference. Prof. Jellinger, do you think the dendritic arborizations in the substantia nigra are pathological markers even in neurons that are not manifesting Lewy bodies or other pathology?

Jellinger: We do not know, and neither do we know what is the role of the pars reticulata of the substantia nigra.

Leigh: If there is dendric atrophy, it may simply be what happens when cells are damaged, because this is well documented in ALS, and this can be a very early change.

Calne: What is the status of the receptors in early stages of PD?

Snow: So far three studies have been done using raclopride: one showed no changes, a second one showed an increase of raclopride binding opposite the lesion, and the last one showed an increase, too. However, these are statistical results that may not be too helpful in an individual case.

Horowski: In PD there is a smaller decrease in HVA than in dopamine levels, so this indicates an enhanced turnover, probably as a compensatory mechanism.

Mizuno

Stern: Is complex I deficiency specific for Parkinson's disease?

Mizuno: Deficit of complex I, from my point of view, is very but not absolutely specific. Shapiro's group studied the activity of complex I in other disorders and they found that it was normal. In my mind, anyway, the deficit of complex I is not the primary event, but a secondary change that I believe is specific for PD.

Youdim: I would like to clarify the topic regarding the platelets: as you know, Parker reported that in the platelets of parkinsonian patients there was a decrease in complex I activity. On the other side, Shapiro did not confirm these data as he found no change in complex I in platelets and muscles. Therefore, we cannot use platelets as a marker.

Riederer: Decreases of complex I and III are very small, according to the literature [10%–30%] and all these enzymes are unsaturated, so what all this means in physiology is unclear to me.

Olanow: All groups confirmed the finding of deficiency in complex I in the nigra. A deficiency of complex I might be among other things associated with a reduced ability to generate glutathione and complex I deficiency could be a consequence of glutathione deficiency. So, complex I deficiency could be secondary to all these other things going on, especially stress induced by peroxidation.

Mizuno: I agree that free radicals could be involved in the degeneration of substantia nigra and in the decrease of complex I. This could be due to a mutation in mitochondrial DNA.

Youdim

Mizuno: Iron accumulation could be a secondary effect due to the degeneration of nigra neurons. In my mind, only very small amounts of iron enter the tissue using transferrin and only a small amount will be eliminated. We do not know how this happens, so the elimination of iron should be studied more extensively. I believe that accumulation of iron is secondary to the energy crisis in the substantia nigra, probably because the elimination of iron from the tissue requires energy.

Youdim: There is a precise period of life when nutrition with iron could cause damage in the CNS. This corresponds to the first 3 years of life in humans and it is related to the formation of the blood–brain barrier. The iron can cause damage in this period and the dopamine system seems to be more sensitive than any other system to iron stress.

Calne: I would stress the point that free radicals are, indeed, able to kill the cells but they are not necessarily involved. It has been demonstrated that the immunological system is able to kill cells too, but this is not necessary either. There are a number of observations that neuroleptics alter the blood–brain barrier causing iron to enter. Now, if iron and free radical formation were the driving cause of cell death, the most disastrous situations should be in the psychiatric hospital where neuroleptics are used extensively. But the evidence is that neuroleptic-induced parkinsonism is usually a transient, dose-dependent phenomenon.

Youdim: It is necessary to reach a critical level of iron to produce the effect, so probably with neuroleptics the level of iron is not sufficient to produce the lesion.

Chapter 18

Alzheimer's Disease

K. Jendroska

Perry

Stern: I think it is important to think about definitions because there is a problem of concepts and categories of definitions leaving aside Alzheimer's disease (AD) or indeed any other disease for the moment. This conference is entitled "Pathogenesis of Neurodegenerative Disorders." It seems to be important at least to separate a description of something in which you attribute certain qualities. You must clearly decide whether you are going to include pathogenic factors—by pathogenesis I understand factors that cause pathology—but it does not necessarily mean that that is synonymous with the dysfunction or disorder that you are studying. So there is a confusion of categories and concepts even before you get off the ground. I think at the very least it is desirable to get a constellation of descriptive things so that we all have a rough idea of what we are talking about. You probably have to separate pathology with a very wide spectrum, because pathology now must include molecular biology. We must make up our mind whether we must include pathogenesis and then there is the separate concept of etiology. Currently we keep wandering in and out of these various things. At least we ought to agree on the problems facing us in concepts and categories. I would like to know what people mean by disorder and by pathogenesis, whether neuropathology is going to come into pathogenesis, whether etiology is a totally different concept, and so forth. Listening to the diplomatically successful presentation I felt wandering in and out of these categories and hence the confusion and the increased liability to misunderstanding and disagreement.

ADVANCES IN RESEARCH ON NEURODEGENERATION, I
D.B. Calne et al.

Perry: I take your point about the use of words here. I don't want to be caught up in that because in a simplistic way I as a neurochemist want someone to tell me that these 10 cases are AD. I want to be certain not to look at a mixed bag of diseases. At this stage simplicity is needed. Unless maybe the molecular geneticist unravels better tools for classification. To be satisfying someone wold have to be able to unravel the whole sequence of events from early maybe environmental influences or genetic ones until we end up with a plaque and a lost neuron. Then you would be able to use these words correctly. But we are dealing with a black box here and maybe this is why we get confused in using these terms here.

Stern: There is one other important implication that occurred in your valiant attempt of definition that has got to lead to failure admittedly from the outset. That is, that definition is dynamic and changes from meeting to meeting I attend. It has to be capable of modification.

Perry: You can say that. But my attempt had led me to take the originial definition people have been more or less satisfied with since we started to do biology on AD and finding that despite molecular genetics one was coming back to a similar definition. We had to come back to the fact that you don't have dementia unless you have plaques and tangles in the cortex. It's no good to define AD as β-amyloidosis because this doesn't distinguish AD from normal aging.

Calne: If the majority of elderly have plaques and tangles then why can't we consider the increase of plaques and tangles the same way as the decrement of raw IQ that occurs between the young and the old and so plaques and tangles and the decrement of psychometric performance is something that occurs in the majority of people, and by definition, if the majority has it then it is normal.

Perry: Here we are dealing with quantitative rather than qualitative differences and left aside whether you call it normal—this is aging.

Calne: If the majority of people over 80 have the plaques and raw IQ scores of less than what they had in their 20s, then isn't that normal? Then where you draw the line to say there are too many plaques or the IQ is too low is something arbitrary. If, for example, you have a Mini Mental Status (MMS) of a certain score someone will call that dementia and someone else will say no. Any time a committee is formed things change and then a new committee is formed and there is nothing absolute about this at all.

McGeer: The problem is people start from normal becoming demented and we are talking about the shade of gray in between.

Finch: Let me ask this question: Suppose we found a case in which there is dense plaques and tangles in the cerebellum—by every definition the neurochemist would say this is the same molecular pathology but it is in the cerebellum. We wold not call this AD because it is in the wrong place. What this reduces to is the pathoclisis, the selective vulnerability that I think is a wonderful conceptual lineage in very specific foci in the brain that can be caused by a number of very different insults. I therefore think that the concept of "alzheimerization" is allowing flexibility to say it can result from extensive β-amyloidosis but it can also result from mysterical changes without deposits that we could recognize. I think this is a fruitful position to be in to focus on the cells and pathways.

Perry: On the other hand you see "alzheimerization" in Parkinson's disease (PD) so that you wonder whether there is AD after all.

Olanow: Clinically we differentiate functional impairment from a disease process. Consider the example of a stroke. A stroke is clearly a disease process but if it is in a "silent area" of the brain the patient is not functionally disabled and in pathology I would say most people of older age have had strokes in their brains. Don't we look at the same thing here? "Alzheimerization" is going on. Under certain circumstances it interferes with cognitive abilities. If you have overwhelming "alzheimerization" it will certainly express functionally but occasionally there may be cases without impairment if critical structures are spared. In other cases minimal changes may cause deficits and herein lies the problem. Some people who have the process are functionally normal; others who are functionally impaired have minimal changes.

Perry: I think that is very useful and we should use definition more in terms of anatomy. We could have "hippocampal alzheimerization," which presents as a memory disorder, and "cerebellar alzheimerization," which might be silent to some degree.

LeWitt: There is a very good example to this concept. It is the disease entity of cortico-basal-ganglionic degeneration. It is a form of degeneration that has a typical unilateral motor and apractic parkinsonian side that evolves into a frontal syndrome and is dementing. There are examples that the same type of pathology is leading to a slowly developing dementia without going through these stages of motor signs. There is a very sparse pathology compared to the amount of cortical dysfunction. Remember that similar syndromes occur in AD patients with initially predominant apractic features or in Hallervorden-Spatz disease. I think the regional definition of these primary progressive disorders is useful for clinicians and might give some regional and pathological clues.

Braak: I believe that we have to use a better definition of "plaque." Many authors consider different things under the name "plaque." Some talk about diffuse amyloid deposits. Here I would propose the term "amyloid deposit" and not the term "plaque." On the other hand, if a plaque is formed of dystrophic neurites and contains intraneuronal abnormally phosphorylated argyrophilic tau protein it is to be called "neuritic plaque." Many authors talk about "senile plaque" and include diffuse amyloid deposits. They easily get 10-fold higher counts than if you restrict yourself to "neuritic plaques." I think it should be made clear by authors what they counted. The other thing is that you proposed AD to be a β-amyloidosis. I would not agree to that. We have seen many cases with mild dementia and mild personality changes that had only intraneuronal changes, neurofibrillary tangles [NFT], and neuritic plaques in the absence of amyloid. Amyloid is by no means a prerequisite for the development of intraneuronal changes. I am not sure what amyloid does in the brain, whether it causes functional deficits. Many authors claim that amyloid is of harm to brain function but this to my knowledge has not yet been shown convincingly.

Perry: Does this mean that you can have a plaque that is not reactive to a β-amyloid antibody?

Braak: Yes.

Jellinger: I remember a recent paper by Probst et al. who have shown that in the very early, the so-called diffuse plaque, there is a decrease of synaptophysin immunoreactivity along the dendrites but there is no definite deposition of APP. Also we have heared from Mesliah and coworkers that the presence of β-amyloid in the very primitive or preplaque is not associated with a significant decrease of synaptophysin-immunoreactivity and synapse count. So I would agree that there is an independent development of cytoskeletal changes, or phosphorylated tau-protein deposits within the neuron, from the deposition of extracellular β-amyloid.

Perry: Don't you find the latest transgenic model very tempting where tangle formation followed β-amyloid deposition in the mouse transfected with a C-terminal APP fragment?

Jellinger: I agree. But we also know that a percentage of mentally normal aged persons have small numbers of tangles in the entorhinal cortex without any β-amyloid deposits. I wonder about the low percentage of about 10% of aged people without AD lesions in your presentation. In other studies I have seen numbers of totally negative cases in terms of AD lesions of 50% to 60%.

Perry: I don't have any better explanation at the moment than geographical differences in successful aging.

Jellinger

Calne: To me the theme of definition recurs. Where does aging end and disease begin? In analogy to, for example, hypertension or diabetes, we see a tendency or trend to occur with aging and at a certain level functional disturbances become manifest. I think you can use the term disorder in that situation and reserve the term disease for the situation that causes the patient's symptoms. A disorder may be present when you reach a deviation of, let me say, two standard deviations of the mean for the given age group. That is all right for hypertension or diabetes. In the brain you have an added complexity. Here you have the effect of lesions in different regions having a different functional expression. I would suggest that in things running in a trend with aging we define something within the range of standard deviations and call that disorder. But if that change is extreme enough or it is in a region of the brain that is eloquent you will have functional deficits, symptoms, or a disease.

Jellinger: I would agree and add that it is not only the quantitative threshold but also time and course. We have seen a number of rather old patients with long-standing dementia and mild AD lesions that lead us to think that the younger people are the more rapid the disease progresses and the more functionally severe the individual becomes affected. We also know from serial biopsies taken over the time of disease that the number of plaques increases and the number of tangles decreases due to the fallout of pathologic changes very large prospective studies will be necessary and you know how difficult this is to do. Admittedly, most of us have only a handful of prospectively followed up cases and this is not sufficient to give us really valid statistical data.

Calne: Another trend I see is that we tend to see the "holy grail" in specific markers and the more we see we learn that they are not specific and that has been holding us back, the idea that there is something that will show us that this is the disease.

Jellinger: I can report on our experience with ubiquitin markers or the Alz-50 in the cerebrospinal fluid that turned out to be absolutely nonspecific and shows no differences between normal, mildly demented, and severely demented patients. The same applies to Lewy bodies or tangles, but also to other intravital markers like amyloid in rectum biopsies, which may indicate systemic disease but are not indicative of dementia or AD.

Horowski: In the last years we have been confronted with the analogy of β-amyloidosis possibly causing AD and impaired dopaminergic transmission as the pathogenetic basis of parkinsoniam. In Parkinson's disease (PD) you can find compensatory mechanisms like striatal dopamine receptor upregulation. The question is whether similar compensatory mechanisms act in the early AD cases you have shown in which you did not find a correlation between mild plaque pathology and dementia. The question is what synaptic markers do if you try to correlate mild changes only.

Jellinger: That we have not yet done but we will take up the suggestion. Coming back to β-amyloid deposition: I think we all agree that this alone does not produce dementia, probably does not interfere with neuronal or synaptic function, nor does it define a disease as much as dopamine dysfunction does not define PD, since you find similar systems affected in totally different syndromes such as progessive supranuclear palsy.

Mizuno: Molecular biology has brought tremendous progress in the understanding of AD. You mentioned that amyloid precursor protein [APP] 770 [amino acid] and APP 695 in the CSF. Normally APP 695 is the predominant form in the brain. In other tissues APP 770 is the predominant form. In Alzheimer brains the relative amount of APP 770 is reported to be increased. This may be due to abnormalities in alternate splicing in AD. When we think about a definition we like to take such factors into account. You showed decrease of APP 695 in the CSF of AD patients that may be due to decreased formation of APP in the brain. I wonder whether there is any reliable method to quantitatively evaluate the APP 695/770 ratio of demented patients *in vivo*. Such a diagnostic test could help establish a diagnosis of AD in living patients.

Jellinger: I reported the results of Beyreuther's group who have found significant differences between normal controls and AD patients in brain tissue APP. I wanted to show this to argue against the idea that these isoforms are derived from systemic APP and to state that they are produced in the CNS tissue. I don't have experience with this particular ELISA test but we have done it on ubiquitin and it turned out not to correlate with the severity of dementia.

LeWitt: What puzzles me as a clinician taking care of AD patients is the fluctuation during the day or under stress, the so-called sun-downing effect. I think that this brings out that there is some cortical functional reserve and that with aging one loses the ability to respond to environmental changes, and once the threshold to clinical disease is reached in the degenerative process the functional reserve gets lost. There is some evidence for that in the increased sensitivity toward anticholinergic drugs. In the aged less anticholinergic produces more deterioration. I wonder whether this reflects some pathological substrate that we could capitalize on that there is some ratio or some algorithm between some cortical output, somatostatin for example, and the number of functional elements lost.

Jellinger: I think we have a very clear pathologic pattern in AD and that those pyramidal cells that are responsible for the cortico-cortical connections are severely depleted and that for instance the primary motor neurons are relatively spared. So that we can state that in the earlier stages we have the fundamental separation of the hippocampus from the rest of the brain in the degeneration of the perforant path but in the very severe stages we have a total neocortical disconnection syndrome that is also evident in the severe atrophy of the anterior two-thirds of the corpus callosum. The distribution pattern of, for instance, tangles in AD is totally different from that in progressive supranuclear palsy. That again stresses differences in local vulnerability in different diseases and may relate to clinical changes. Unfortunately for us, cases coming to post mortem are usually late stages due to the fact that the patients survive for a long time. The very mildly demented are now the major target group for research that will give important data about the earlier stages. This is where we need more research.

Leigh: Let mc make the comment that pathologists have almost nothing to contribute to a discussion about the mechanisms of a disease. Looking for example to the prion disorders where we know that the pathology is caused by a specific molecular change and we see that the pathological picture varies from the so-called classic spongiform encephalopathy to the almost nonexistent. And therefore we have to hand the baton on to the cell biologists at a certain stage and say we know that there is a mutation at a certain gene, let's say in familial AD, but there are a dozen ways how this could interact with genes or environmental influences to produce tangles or whatever. We accept that in a group β-amyloidosis may be the most correct designation we have at the moment but that may not account for all cases of that pathology and that pathology may not tell us much about the biochemistry.

Jellinger: I fully agree. We have the same situation with chromosomal disorders and developmental changes in the brain that may show from anencephalia to holoprosencephalia to very mild changes maybe even only cellular deviations with the same chromosomal defect. And here we may have the same situation. This question has to be asked to the molecular biologists.

Riederer: I am irritated by the fact that β-amyloid doesn't correlate well with dementia and plaques and tangles in the transgenic mouse don't correlate with anything like dementia. Could you make a comment to the value of these features? Are they markers or are they not markers?

Jellinger: This is an important point here. The correlation exists only for the very demented, the burned out cases where we really see a high increase of these so-called markers. This holds not true for the early stages of the disease. It is not valid for the mildly and moderately demented patients. In the so-called mildly demented patients with a Mini Mental Status score around 28 we may find nothing at all—no β-amyloid, no plaques, no tangles. We may find comparatively large numbers of tangles in the entorhinal cortex, we may also find a moderate number of primitive plaques and few tangles in the isocortex. All these neuropathological patterns may clinically occur as mild AD. I remember the discussion with Hornykiewicz who wanted to do biochemical studies on PD without neuropathological diagnosis. He was convinced. The same is true for aged persons and AD. Dementia doesn't mean someone had AD, nor do plaques and tangles mean someone has AD. We really have to take all parameters that we can get hold of into account.

Braak: The clinical fluctuation mentioned earlier may be reflected by a rather enigmatic change in the suprachiasmatic nucleus where in AD we see a rather serious neuronal fallout not caused by tangles and there is also no β-amyloid pathology. The suprachiasmatic nucleus influences the circadian rhythm. It controls the production of melatonin and it may well be that the fluctuations in AD are caused by changes in hypothalamic functions not taken into account by earlier pathologists using the usual diagnostic methods.

Jellinger: In most recent studies the neuropsychological examination has been done every 6 months. In the older studies 12-month or 24-month intervals were commonly used. In some cases of AD the progression especially in the later stages is very rapid and so the older correlative studies are not quite valid. You have to do neuropsychologic evaluations under standardized conditions at the same time of the day, strictly avoiding acute decompensating influences like common cold or pneumonia.

McGeer

Jellinger: Would you call AD an autoimmune disease of the brain?

McGeer: I would term it autodestructive because most people understand autoimmune as getting a clonal production of antibodies against a normal protein. This may not be the case. It just may be that the phagocytes are ravaging everything.

Jellinger: As you know, Wisniewski is working on the hypothesis that the β-amyloid is produced or processed in the microglia and a number of ultrastructural studies have been presented to show that the macrophages behave differently in AD. He showed an infarction and the phagocytes did phagocytosis also of the β-amyloid but there were peculiarities leading to the hypothesis that β-amyloid might be produced by the microglia. But if I understand you correctly, you may change this concept in that there is probably a deposition of some potentially harmful substance against which the brain cells are fighting.

McGeer: We have just finished a study in which we were able to show all segments of APP in microglia but the only sequence of APP that occurred extracellularly was β-amyloid. If you look for the APP message in the area of the plaque by *in situ* hybridization you don't find it. It is known that the microglia can make APP but the critical test is whether it actually is making APP and it has been shown by Dr. Finch that it doesn't.

Poewe: I would like to take up two points. First, I was surprised that the diagnostic accuracy in AD that you have presented in your data can be quite confidently made by clinicians. The amount of correct diagnoses in your referral centers is much higher than what we are used to see in Parkinson's disease. The UK Parkinson's Disease Society Brain Bank reports an error rate of well about 10% even within the referrals from highly specialized centers. Leaving aside the more difficult diagnoses like Braak's disease, the number of correct diagnoses is quite consoling, especially in light of our earlier discussion on the difficulties of defining AD. Second, I found the aspect of lateralization you mentioned in your talk very interesting because it highlights another element of pathoclisis that we have not much addressed yet. In PD we are used to the observation of lateralization and even considering it a diagnostic prerequisite. Maybe there is more lateralization in AD than one knows from pathological studies. I take it that usually one-half of the brain goes to autopsy and the other half is used for biochemical studies.

McGeer: Twenty-two of our 26 studied cases clearly had lateralization to the dominant hemisphere. It may be that those who had it on the other side didn't show up in the clinic. But by the time we look at the brain post mortem it is just filled with plaques and tangles. I personally was amazed how you could take dementia and there are tangles, 85% of the time. The other thing I was amazed of was how little multi-infarct dementia there was. Even if there is some it can easily be picked up by CT scanning. Some of the people who had infarcts also had amyloid angiopathy so that there was mixed dementia in this situation, which clearly predilects to vascular accidents. That even enforces the rule: Call it Alzheimer and you will be right most of the time.

Poewe: I think this is an important point and I still don't understand how it came into clinics to say people with multiple infarcts have multi-infarct dementia and not AD.

Jellinger: I want to stress a point concerning positron emission tomography [PET] in AD. Usually we find a more severe neuronal loss in the frontal and in the temoral lobe than in the parietal lobe, though as you showed in your PET data the parietal lobe suffers more in terms of depressed metabolic turnover, which probably reflects a more severe loss of synapses and dendrites than of neurons compared to the other regions. Second, the usual accuracy rate of clinical diagnosis to post-mortem morphologic diagnosis in modern studies ranges between 82 and up to 100%, though admittedly the 100% have been found in two very small American studies in highly selected material. Therefore, we suggest an average accuracy rate of clinical diagnosis with modern test tools of about 85%. I think we should not rely too much on the Hachinsky Ischemia Scale, which in our hand gave wrong results in about 30%. We had too high Hachinsky scores in AD patients. We have seen small infarcts and lacunar states in about a third of cases with classical AD, which we should not classify as mixed dementia. We should use the term mixed dementia only in those cases that have a true combination of AD and multi-infarct dementia. That means at least a 50- to 100-cm^3 tissue loss due to infarcts. I think this is not strictly observed in all studies.

Stern: Would you please comment on 1% decrement a month of glucose metabolism that you have seen in you PET study? Does it mean 12% a year and after 9 years there is no more brain glucose left?

McGeer: We don't know what the endpoint of AD in PET is. There is a window in which it is convenient to use PET as a tool, that is, when you can be reasonably certain that a person has enough dementia that justifies a diagnosis of probable AD. There they are already several years into their disease. At that stage they continue to deteriorate and you may have the opportunity for another PET and then they are not mentally able to cooperate anymore. And there is this relationship to parkinsonism in AD and the people are getting so rigid that they cannot be submitted to PET anymore.

Calne: Please make clear to me 1% of what it is that gets lost. Is it 1% of normal peoples' glucose values or 1% of an initial value that already must have been very low?

McGeer: The average decrement in our 20 serial scans we did worked out to be a little over 1% per month of the initial scan, which indeed is already way below normal.

Finch: I want to come back to your argument on the use of nonsteriodal anti-inflammatory drugs. Can you give us any information as to the clinical course of people treated with these drugs?

McGeer: Only that we found that dementia scores of people with arthritis is only one-tenth of that of the general population. There were selective factors in our study though, and it stays controversial. There is no prospective study so far. There is the Dapson study where people were in the same institution and there were significant differences between the Dapson treated and the non–Dapson treated.

Horowski: I have some 20-year-old experience from taking care of patients with Down syndrome. They obviously deteriorated mentally but the cause of death in most cases apart from cardiovascular complications was due to tuberculosis or bronchopneumonia from an underlying defect of the macrophage-monocyte system in these patients, or there was acute leukemia. I wonder whether there is any known link between these observations and the defects in the immune system that you have reported for AD.

McGeer: We all know that before the era of antibiotics most children with Trisomy 21 died early from infection. Two things occur in Trisomy 21 in the obligatory region of Down syndrome. First there is the interferon receptor and second there is the lymphocyte function associated antigen 1, LFA 1, so that an extra dose of either of these genes will cause trouble in the immune system, though I can hardly imagine how this would relate to AD.

Dalakis: I wonder whether you could elaborate on your data on the autodestruction. I have the following reservations about your findings. First, the presence of macrophages does not necessarily indicate an autodestructive process. They may just clear debris. Second, the presence of major histocompatibility complex 1 [MHC 1] or MHC 2 in the area of destruction is to be expected. You are going to see MHC 1 or 2 in the normal cells surrounding the lesions. Third, to the presence of complement. We know from Duchenne's muscular dystrophy, which is certainly not an autoimmune disease, that necrotic fibers display membranolytic attack complex [MAC]. The complement complex may be involved as part of the necrosis but not as part of an autoimmune process. The last question I have is, are the lymphocytes you found CD4 or CD8? Because if they are CD4 they could be just macrophages because the antibody against CD4 recognizes macrophages.

McGeer: First of all, both CD4 and CD8 are there. Second, once you start generating complement of course it is going to go on to the formation of membrane attack complex. The membrane attack complex is there to knock out cells. That is its function and we want it to kill bacteria. So if the MAC is there and it's inserted into a cell membrane it will kill that cell. It would go into the cell membrane unless the cell were own. The reason why you have these defensive proteins in all the cells of your body is to prevent this very thing happening. So I would take it—whether in AD or in muscular dystrophy—that if you have an upregulation of these defensive proteins you have got a cell that is exposed to the MAC and still has the capacity to transcribe these genes to make extra amounts of the protective proteins. In the case of AD—I can't comment on Duchenne's—we found five proteins that colocalize with damaged neurons, all of which are there to defend against the various complement proteins including the MAC. And you see the MAC decorating dystrophic neurites, tangled neurons, and neuropil threads. We almost got a complete colocalization.

Dalakis: What about MHC antigens in the surrounding healthy cells?

McGeer: MHC class 1 are vigorously expressed by capillaries. It is expressed at a low level in humans and at a high level in rats on activated microglia and you see very rare but huge astrocytes that also carry MHC 1. MHC 2 is more specific to activated microglial cells. When the microglia tool up to become phagocytic that is when it is throwing out its MHC 2 antigen on its surface. And this is why it is such an excellent marker in neuropathology.

Olanow: I want to add to the discussion that MRI scanning also has been used to identify AD showing some atrophy of the medial temporal lobe on coronal sections. There has been a fairly high correlation with autopsy-proven cases of AD. The advantages of this technique is that it is available in almost all centers. And again a unilateral predisposition was observed. I must mention that as something that complements your findings with PET. One thing I was hoping to hear following the discussion was some unifying hypothesis in pathogenesis that would explain why cells die regardless of which form of expression neurodegeneration takes. Maybe it is too general a question but there is some evidence of autoimmune phenomena in PD, some dopamine antibodies have been found. There has been some evidence of autoimmune phenomena in ALS as well. Do you think this might be something that underlies degeneration, and having said that, how can you know that what you see is not simply a response to dead material generating an immune reaction?

McGeer: Again, the only evidence that we could come up with that differentiates between the scavenging of debris and the active attack is the presence of the MAC and the protective proteins that defend against this. But of course we do see these changes in other diseases that you described. So regardless of specific antibodies, if you start a microglial cell on a rampage it may do more than just clean up debris. That is the credo of the "Non-SAIDists" [non-steriodal anti-inflammatory drugs]. That it is the microglial cells that can't get their jobs done, trying harder and harder and pretty soon start spitting out deadly juices that overcome the ability of the neuron to defend itself.

Riederer: Would you suggest the use of corticosteriods in the treatment of AD?

McGeer: We initially have thought about that. We put some patients on steroids. There were huge differences between the steroid group and the indomethacin group. Patients on indomethacin held their own whereas patients on steroids got worse. Steroids themselves have potent actions on neurons. Neurons have steroid receptors and I think that there are many other things steroids do in addition to simply suppressing general inflammatory reaction. It is possible that steroids by themselves are not really harmful but the differences in the groups were really very striking.

Finch: Sternberger and others have argued that plasma proteins may leak into the brain tissue during the death process, which might allow complement to form. To my knowledge no biopsy material has been studied yet that would have shown that complement complexes are formed in earlier stages of the disease. This I think would be an important work to do.

Braak

Finch: At the cellular level in the transentorhinal cortex do all of the neurons eventually show the pathologic changes?

Braak: It's all of them, in particular in the transentorhinal region, which is the first area to be involved. In the late stages of AD you will fail to find this layer. It has completely disappeared. So this is another point. If you count tangles you must be aware of the fact that even the tangle goes away. The tangle has its own biography. It is born in a nerve cell as a tiny compound of a lipofuscin deposit. Then it forms a large tangle, then it becomes a ghost tangle, later it gets under attack of glial cells, and years later it is gone.

Finch: I think this is important because it rules out a class of theories about the cellular disorder that might otherwise be considered as a multi-step process like the acquisition of malignancy. This rules it out because it affects all cells of a given type.

Braak: It is very specific and this is why I as an anatomist like to study AD. It is not a diffuse process. It affects only a few cell types in the brain.

Perry: I am not sure on what you base this staging. Did you look at normal people and added them to AD?

Braak: I did not rely on a diagnosis of AD. I have studied the normal range of brains I obtained from pathology. The transentorhinal changes are only the beginning of structural alterations. They are normal. I don't imply cognitive decline. It's just the beginning in terms of morphology.

Perry: I looked at many transmitter receptors in the hippocampus and the adjacent gyrus, and as I mentioned, most of them are glutamate receptors of CA 1. But you can find nicotinic receptors in the hippocampus. They suddenly appear in the presubiculum and right into the entorhinal cortex and this stands out compared with all other brain areas. And it is striking that what you describe as the area that first gets the changes has this nicotinic receptor on it. In PD nicotinic receptors are concentrated in nigral neurons. The interest in PD lies in the tempting correlation between smoking and the protection against degeneration—a different pattern than what is seen in AD where it is somehow involved in cell death.

Braak: We don't have enough data on these pyramidal cells but they seem to be glutamatergic. The perforant path is glutamatergic. I don't know why they are so different to other pyramidal cells generating glutamatergic fibers; for instance, the next link of the chain behaves totally differently. The perforant path goes up to the fascia dentata of the dentate gyrus. The next link, the mossy fiber system of the fascia dentata to the CA 3, is extremely well preserved in AD. There are striking differences between the perforant path and the next link the mossy fiber system. I don't know whether zink as a component of the mossy fiber system plays a role as a defensive element of that system.

Finch: Zink concentration in AD doesn't change and it stands out as a marker because its system is well preserved.

Jellinger: According to a recent study by Hoff et al., there is a significant difference to PSP in that in PSP the neurofibrillary tangles are mainly in the granular cell layer of the dentate gyrus similar to Pick's disease, which speaks for a selective vulnerability in different cell types in these different disorders.

Ransmayer: Did you quantify cell loss in the basal nucleus of Meynert in your cases?

Braak: Nucleus basalis is affected secondarily. It is affected relatively late in the course and follows the limbic parts of the thalamus. In general, the subcortical nuclei show much more variation in the sequence of degeneration compared to the cortical regions.

Jellinger: I think that there is a striking difference between AD and PD in that most of the subcortical changes in AD seem secondary to damage of their cortical target areas, which is something that we almost never see in PD. I would hypothesize that subcortical changes in AD are secondary to cortical changes, whereas in PD there is a more or less parallel course of both cortical and subcortical changes.

Braak: But it is interesting that demented PD patients do show Alzheimer pathology. Almost all of the cases we have studied showed stage III or IV. And interestingly, the process in PD seems to arrest there. So the most clear-cut cases of entorhinal affection we have seen in PD. There is a tendency in PD not to develop much isocortical change.

Jellinger: Unless they are severely demented. You know that we have recently shown in a series of PD that there is a good correlation between the Mini Mental Status and the severity of AD lesions. So we think that in demented PD patients the PD changes are superimposed by AD lesions and we found almost the same course and distribution pattern as in classical AD.

Braak: I think we have to consider giving up the concept of subcortical dementia.

McGeer: I would confirm. We have not found a case that we would call subcortical dementia. People have recently examined the frequency of AD-type dementia in parkinsonians versus their spouses. It is something like a ninefold higher incidence of dementia in the parkinsonians. So there is really a link between dementia and PD.

Jellinger: Let me add something to the concept of subcortical dementia. To my opinion it exists in a few types of lesions: primary thalamic degeneration, which is extremely rare and may be associated with Creutzfeldt-Jakob disease, cases of Binswanger's disease with exclusive damage to basal ganglia and white matter, and in some cases of leukodystrophy. The term is certainly not appropriate for PSP, which is a frontal type of dementia, and also not appropriate for Huntington's disease and PD, which were the three for which the term has so often been used.

Olanow: Clinically we think that 30% of PD patients develop dementia in the course of their disease. This percentage is much higher if you look at patients with late-onset PD. We have been interested in it because we consider development of dementia a very valuable endpoint for clinical trials for neuroprotective drugs, especially compared with such soft endpoints like "time to levodopa," which has been used in the DATATOP study. What do you think about Lewy body dementia? We have heard so much about Alzheimer dementia in PD.

Braak: We have followed that up in our cases. They had nigral but no cortical Lewy bodies.

Jellinger: In our consecutive autopsy series of 26 dementia cases we have one with incidental Lewy bodies in the cortex but not diffuse Lewy body dementia [DLBD]. And in a different consecutive autopsy series of 25 PD patients we had no case with DLBD. There is our older study of 300 cases of PD. Among the younger ones dying at age 60 to 65 years we had 18% with additional AD pathology, whereas in the aged PD group dying around 80, we found AD lesions in 84%. So we suggest that in about 35% to 50% of PD there is sufficient AD pathology to allow a diagnosis of superimposed AD.

Braak: I would like to get a consensus of the pronunciation of Lewy [Dr. Braak pronounces German]. Lewy was a high-ranked military neurologist in Germany during the first world war. Being forced to leave Germany he became a high-ranked military neurologist for the Americans in the second world war. I always feel it inappropriate to pronounce it American.

Jellinger: His name was Friedrich Wilhelm Lewy [Dr. Jellinger also pronounces German].

Perry: It is an old question whether it is PD that leads to cortical Lewy bodies [pronounced German]. Is it AD with concomitant Lewy bodies [pronounced American]? I think it's unresolved. To my own opinion in older parkinsonian patients you can get increased "alzheimerization" and dementia. For epidemiological studies it needs to be resolved because if you call parkinsonistic AD PD you get a much larger group to follow up.

Calne: To my opinion the whole concept of DLBD is an ill-chosen terminology. One is dealing with a pathological marker that can occur in many different places, presumably indicating a similar kind of change just taking place in different parts of the brain. And if the changes are found primarily in the cortex you call it dementia, if it's in the nigra it's PD. And what you're dealing with is just the focality of the lesion. Let me ask: in the cases of DLBD are there plaques and tangles as well?

Perry: They are in some. But we have about 20% of cases with no Alzheimer lesions looking otherwise like a normal elderly individual.

Jellinger: What is your interpretation of the basis of dementia in these cases without AD lesions?

Perry: I think the defect is in the cholinergic system.

Jellinger: I cannot agree to the suggestion that there is a correlation between the number of Lewy bodies in the cortex and the severity of dementia as some do. Another thing that we have to consider is that the majority of the parkinsonian patients are getting older. In our first study at the early stages of the L-DOPA era the mean age at death was around 62. In our recent series, which was admittedly selected because the patients came from geriatric hospitals, the mean age at death was around 83. So they are two decades older and this may in part explain the increasing superposition of AD lesions. It is certainly not the only reason. Another point is the increasing incidence of psychiatric complications, psychosis, and confusion. They were much lower in the pre–L-DOPA era, but when we did morphological studies we did not find any explanation for this and therefore we think this is a drug-induced phenomenon, probably L-DOPA, and not a feature represented by morphological or neurochemical changes.

Riederer: Which morphological changes do represent schizophrenia compared to AD? Both have been related to changes in the entorhinal cortex. Is there is possible explanation why demented or parkinsonian patients are more susceptible to psychosis or pharmacotoxic psychosis?

Braak: In all cases of dementia we found complete destruction of layer prealpha in the entorhinal cortex. The neurons are gone. In schizophrenia there are many indications to abnormalities of neuronal localization. They do not reach their target areas, neurons are seen in clusters, and so on.

Riederer: If these cells are glutamatergic then in both cases, AD and schizophrenia, it may be possible that there are similar functional defects in the perforant path. If this is the case you would expect a predominant dopaminergic tone in the entorhinal region that would fit with the dopamine hypothesis of schizophrenia and also pharmacotoxic psychosis.

Olanow: Let me come back to the so-called markers of neurodegeneration. Do you think they represent the same process just in different cell types? I speak of Lewy bodies and tangles. They both contain high amounts of aluminum and iron. The origin of tangles in lipofuscin raises the question of oxidative stress as a pathogenic factor similar to what is discussed in PD pathogenesis. Would you go so far to say that oxidative stress may be the common causative pathogenic factor underlying these neurodegenerative processes causing this marker in this cell type and that marker in the other?

Braak: I would not like to make too simplified statements. But it seems important that there are different cell types involved. The cortical neurons show a wide spectrum of cell types and are subjected to very different conditions.

Calne: But it is filamentous cytoskeletal material that Lewy bodies and tangles are made of. This is a unifying theme.

Jellinger: We did not find increased iron in Lewy body by x-ray analysis. But we found it to be increased in the neuromelanin. A recent study from Paris did not confirm the increase of aluminum in the neurofibrillary tangle and I think we must be very cautious to relate these different cytoskeletal changes to each other.

Olanow: I think there is much controversy in this area of research and you very much have to consider the different levels of sensitivity of the various methods comparing for example results from x-ray analysis with LAMA. I think one reason why our literature is so full of contradictions lies in the variability of sensitivity of all those methods. Using LAMA, for example, we indeed found increased iron in the Lewy bodies.

Youdim: I think that with methods with such a high sensitivity you are measuring insignificant amounts that don't bear any importance. You find iron everywhere, in every membrane and so on.

Finch: How do you compare Down syndrome morphologically with AD?

Braak: Both follow quite the same sequence. There is basically the same morphological picture at a younger age in Down syndrome compared to Alzheimer's disease.

Chapter 19

Pathogenesis of Neurodegenerative Disorders

G. Ransmayr

Meeting chaired by W. Olanow

Motor neuron disease (MND) consists of a group of paretic syndromes of heterogenic presentation and clinical course involving the perikarya of lower motor neurons. Amyotrophic lateral sclerosis (ALS) (Charcot, 1877) is the most typical and frequent form of MND presenting with symptoms of the anterior horn cells (multifocal weakness, fasciculations, cramps, muscular atrophies, and loss of tendon reflexes) and corticospinal tract lesions (spasticity, hyperreflexia, and pyramidal tract signs). Involvement of the bulbar cranial nerves, the cervical spinal cord, and the corticobulbar tract leads to dysarthria, dysphagia, and respiratory failure (Adams and Victor, 1989). Loss of emotional control occurs as an element of pseudobulbar palsy (Ziegler, 1930). Progressive spinal muscular atrophy (PSMA) and progressive bulbar palsy (PBP) are less frequent forms of MND characterized clinically by purely lower motor neuron involvement and rapid progression. However, degenerative changes may also be found in the pyramidal tracts (*Calne, Mizuno, Jellinger*; Rowland, 1982). Whether or not primary lateral sclerosis (PLS) (*Calne*; Erb, 1902; Younger et al., 1988) and chronic spinobulbar spasticity (Gastaut et al., 1988), two rare diseases with purely upper motor neuron symptomatology and less malignant course, represent subsets of ALS remains uncertain. Around 5% to 10% of ALS cases are familial and 90% to 95% sporadic (Calne; Mulder, 1957).

Clinical diagnosis of MND is substantiated by characteristic EMG findings (widespread fibrillations, positive waves, fasciculations, chronic

ADVANCES IN RESEARCH ON NEURODEGENERATION, I
D. B. Calne et al., Editors

neurogenic changes, giant potentials, and loss of motor unit recruitment), absence of craniocervical malformations, compression of the cervical spinal cord or cervical or lumbosacral nerve root compression, exclusion of a neoplastic process [Hodgkin's and Non-Hodgkin's lymphoma (*Jellinger*; Younger et al, 1991), multiple myeloma, carcinoma], paraproteinemia, heavy metal intoxications, metabolic disorders (diabetes, vitamin B12 deficiency, adult hexosaminidase A and B deficiency), and persistent viral infections (for review Mitsumoto et al., 1988). There is general agreement to Dr. Leigh's proposal that an operational definition of ALS based on clinical grounds is needed (possible, probable, and definite ALS) to improve diagnosis, classification, and insight into evolution of MND. Even though the clinical presentation of MND deserves further accurate evaluation, the reliability of clinical diagnosis of ALS and PLS is around 95% , which is higher than that of Parkinson's disease (80%) and dementia of the Alzheimer type (65% – 86%); *Olanow*; see also Rowland et al., 1991; Younger et al., 1988; Hughes et al., 1992; Tierney et al., 1988). The markedly higher prevalence of Parkinson's disease in patients with ALS as compared to controls deserves clinicopathological correlative quantitative assessment of substantia nigra dopaminergic neurons in ALS (*Poewe*). One also should consider the possibility of an overlapping syndrome (*Olanow*). Despite the absence of clinical features of parkinsonism in ALS-parkinsonism-dementia complex of Guam, there may be striking loss of substantia nigra neurons at autopsy. Parkinsonian symptoms could be concealed by upper and lower motor neuron symptoms or would probably evolve if the patients lived longer.

In Parkinson's disease and Alzheimer's disease changes in the Edinger-Westphal nucleus are found (*Jellinger*). Motor neurons of the cranial nerves III, IV, and VI and the spinal Onuf's nucleus (sphincter control) are rarely affected in MND (Hirano et al., 1969; Mannen et al., 1982). Impairment of rapid eye movements and upward and downward gaze can be found in late stages of ALS, resulting rather from degeneration than from terminal hypoxia (*Olanow, Mizuno*). This assumption was confirmed by Leigh et al., who visualized ubiquitin-immunoreactive inclusions in the third nerve nuclei of patients with advanced ALS. Ubiquitin-immunoreactive inclusions in anterior horn cells (Lewy — body-like inclusions, Bunina bodies, Bunina bodies-related structures, and meshwork of fine filament) are characteristic for ALS and light microscopically and ultrastructurally different from ubiquitin-positive inclusions in Parkinson's and Alzheimer's disease (*Jellinger*; see also Murayama et al., 1990; Leigh et al., 1988).

Various environmental factors and genetic disposition could underly ALS (for review, see Mitsumoto et al., 1988). It is hypothesized that inflammatory processes could play a role in the pathogenesis of ALS (*McGeer*). Functional proteins involved in inflammatory processes, however, cannot be analyzed in formalin-fixed and paraffin-embedded material (*McGeer*). Inflammatory processes and immunoglobulin staining were described on the surface of anterior horn cells of ALS patients as well as passive transfer (*Olanow; Dalakas*). Antibodies directed to gangliosides of peripheral nerves occur in 50% to 75% of patients with ALS and IgM deposits were found to bind to lower motor neuron structures (*Dalakas*; Thomas et al., 1990; Mitsumoto et al., 1988, Drachman and Kuncl, 1989, for review). In patients with "ALS-like syndrome" attributable to monoclonal gammopathies, IgM paraproteins were found to react with gangliosides of the motor neuron (*Dalakas*; Latov et al., 1988), which corroborates the hypothesized role of autoimmune processes in the pathogenesis of "ALS-like syndromes."

In a Japanese family linkage of MND to chromosome 21 was revealed (*Mizuno*). No linkage of MND to chromosome 21, but positive lod scores were recently described by Siddique et al., 1990). In Kennedy's syndrome (X-linked bulbospinal muscular atrophy) genetic linkage studies indicated a proximity of the gene for the disease and the androgen receptor (Kennedy et al., 1968; Warner et al., 1990). This finding might serve as a clue to explain the preponderance of MND in men. The abnormality of the androgen receptor in Kennedy's syndrome is an increased size of the polyglutamine repeat sequence in exon 1 (*Leigh*). Western blot analyses of the abnormal androgen receptor have presumably so far not been published (*Leigh, Mizuno*). The androgen receptor is expressed in anterior horn cells, however, its function in the anterior horn cell is unclear (*Leigh*). Sexually dimorphic motor neurons like the Onuf's nucleus might be regulated by androgens, as demonstrated in rodents. The trophic effect of testosterone on motor neurons supplying the bulbocavernous muscle (homologue of Onuf's nucleus) in the rat may be mediated by testosterone-sensitive trophic factors originating from the muscle (Arkai et al., 1991), whereas other muscles are testosterone-insensitive. One could speculate that this muscle-derived neurotrophic factor is preserved in the perineal muscles in MND and lost or deficient in other muscles (*Leigh*). However, the motor neurons of Onuf's nucleus are preserved in Kennedy's syndrome (*Leigh*; Sobue et al., 1989). The molecular abnormality in Kennedy's syndrome may be related to abnormalities to the gene regulation (*Leigh*).

There is little knowledge concerning monozygotic and dizygotic twins in ALS (*Leigh*). As in familial Alzheimer's disease, there may be several genetic abnormalities in familial ALS. Identification of these changes probably will yield fundamental information on the interaction of genetic abnormalities and environmental factors that produce the phenotype. Multiple mutations of the prion gene give rise to disorders with a quite variable phenotype. Polymorphism at the methionine-valine site seems to be a susceptibility factor. Some patients with ALS-dementia complex might have prion disease, although there is no direct evidence for this notion. Leigh et al. have studied the brain of one patient showing showing the typical features of ALS-dementia as described by Neary et al., 1990. The sister of the patient dies from an illness lasting only a few months. The brain showed the neuropathological characteristics of spongiform encephalopathy of Creutzfeldt-Jacob type. The father of both patients also had a dementing illness (*Leigh*).

In ALS, as in Parkinson's disease, most patients are sporadic. Some patients suffer from an autosomal dominantly inherited disease (*Mizuno*). The genetic predisposition in Parkinson's disease is rarely so extreme that trivial or common environmental stimuli produce clinical disease and positive family history. As positive family history is rare, one should assume only minor genetic predisposition or the environmental factor is rare, its dosage not critical, or the environmental factor wouls have to occur at a crucial moment during life. Subclinical Parkinson's disease could be particularly common in the family of someone who has obvious Parkinson's disease. The same argument can be made for ALS. Looking at factors that control hypersusceptibility to chemicals in different strains of laboratory animals, one may learn much about differential predisposition to toxins (*Williams*). Correspondingly, among several members of one family having consumed MPTP, only one developed parkinsonian syndrome (*Calne*). However, if the interaction of the gene and the environment would be that important, one would expect clusters in epidemiology. However, the pattern of epidemiology of Parkinson's disease, as in ALS, is not cluster. The question arises whether the exposure of the body to a toxin is one-time or ongoing (*Olanow*). The underlying mechanisms of exposure are speculative. Parkinson's disease may be a multistage process with different mechanisms at different stages of the disease or a long-term exposure to a toxin. Alternatively, it could be similar to renal failure; once a certain level of kidney damage is reached, the patient is destined to go into renal failure, whatever the cause and even if the cause is removed (*Williams*).

There are some reports in the literature of classical ALS developing after paralytic poliomyelitis (Dalakas et al., 1986; Armon et al., 1990). The progression of weakness in acute polio is not extremely quick, progressing from one limb to another over a certain period of 3 to 5 days and reaching the peak of paralysis after 7 to 10 days (*Dalakas*; Jubelt and Lipton, 1989). Acute paralytic poliomyleitis is defined as a disease caused not by just the poliovirus, but all the enteroviruses (*Dalakas*). Acute paralytic polio is a multifocal, but generalized, disease. Almost all postpolio patients have focal disease even within the same limb, which is probably related to the "threshold phenomenon" of the number of the damaged neurons (*Dalakas*). In tick-born encephalitis there is also focal involvement, (e.g., in the proximal part of the limb) (*Jellinger*). Approximately 60% to 70% of the motor neurons in a spinal cord segment have to be destroyed before paralysis of the muscle supplied by the segment is clinically noticeable. Certain segments of spinal cord are more affected than others, and subclinical disease is widespread (*Dalakas*). Neuronal death is, at least under experimental conditions, a rapid event and does not take more than a number of hours. But when the energy sources of neurons are compromised, which can be a prolonged state, and a second event occurs, then the time course is very important (*Finch*).

References

Adams RD, Victor M (1989): *Principles of Neurology*, 4th ed., New York: McGraw-Hill.

Araki I, Harada Y, Kuno M (1991): Target-dependent hormonal control of neurone size in the rat spinal nucleus of the bulbocavernosus. *J Neurosci* 10:3025–3033.

Armon C, Daube JR, Windebank AJ, Kurland LT (1990): How frequently does classic amyotrophic lateral sclerosis develop in survivors of poliomyelitis? *Neurology* 40:172–174.

Charcot J-M (1877): *Lecons sur les Maladies du Système Nerveux*. Paris: Delahage et Lecrosnier.

Dalakas MC, Elder G, Hallett M, Ravits J, Baker M, Papadopoulos N, Albrecht P, Sever J (1986): A long-term follow-up study of patients with postpoliomyelitis neuromuscular symptoms. *N Engl J Med* 314:959–963.

Drachman DB, Kuncl RW (1989): Amyotrophic lateral sclerosis: an unconventional autoimmune disease? *Ann Neurol* 26:269–274.

Erb WA (1902): Spastic and syphilitic spinal paralysis. *Lancet* 2:269–274.

Gastaut JL, Michel B, Figarella-Branger D, Somma-Mauvais H (1988): Chronic progressive spinobulbar spasticity. A rare form of primary lateral sclerosis. *Arch Neurol* 45:509–513.

Hirano A, Malamud N, Kurland L, Zimmerman HM (1969): A review of the pathologic findings in amyotrophic lateral sclerosis. In: *Motor Neuron Diseases*, Norris FH Jr, Kurland LT, eds. New York: Grune and Stratton.

Hughes AJ, Daniel SE, Kilford L, Lees AJ (1992): Accuracy of clinical diagnosis of idiopathic Parkinson's disease: a clinico-pathological study of 100 cases. *J neurol Neurosurg Psychiatry* 55:181–184.

Jubelt B, Lipton HL (1989): Enterovirus infections: In: *Handbook of Clinical Neurology*, Vol. 12(56), Vinken PJ, Bruym GW, Klawans HL, McKendall RR, eds. Amsterdam–New York: Elsevier Science Publishers.

Kennedy WR, Alter M, Sung JH (1968): Progressive proximal spinal and bulbar muscular atrophy of late onset. A sex-linked recessive trait. *Neurology* 18:671–680.

Latov N, Hays AP, Donofrio PD (1988): Monoclonal lgM with unique specificity to gangliosides GM_1 and GD_{1b} and to lacto-N-tetraose associated with human motor neuron disease. *Neurology* 38:763–768.

Leigh PN, Anderton BH, Dodson A, Gallo J-M, Swash M, Power DM (1988): Ubiquitin deposits in anterior horn cells in motor neurone disease. *Neurosci Lett* 93:197–203.

Mannen T, Iwata M, Toyokura Y, Nagashima K (1982): The Onuf's nucleus and the external anal sphincter muscles in amyotrophic lateral sclerosis and the Shy-Frager syndrome. *Arch Neuropathol* 58:255–260.

Mitsumoto H, Hanson MR, Chad DA (1988): Amyotrophic lateral sclerosis. Recent advances in pathogenesis and therapeutic trails. *Arch Neurol* 45:189–202.

Murder DW (1957): The clinical syndrome of ALS. *Mayo Clin Proc* 32:427–436.

Murayama S, Mori H, Ihara Y, Bouldin TW, Suzuki K, Tomonaga M (1990): Immunocytochemical and ultrastructural studies of lower motor neurons in amyotrophic lateral sclerosis. *Ann Neurol* 27:137–148.

Neary D, Snowden S, Mann DMA, Northen B, Goulding PJ, Macdermott N (1990): Frontal lobe dementia and motor neuron disease. *J. Neurosurg Psychiatry* 53:23–32.

Rowland LP (1982): Diverse forms of motor neuron diseases. In: *Human Motor Nueron Diseases*, Vol. 36, Rowland LP, ed. New York: Raven Press.

Rowland LP, Santoro M, Lange DJ, Hays AP, Thomas F, Fink ME, Wadia N, Latov N (1991): Diagnosis of ALS. *Ann Neurol* 30:225–227.

Siddique T, Pericak-Vance MA, Roos RP, Watkins P, Hung W-Y, Brooks BR, Noore F, Tandan R, Nicholson G, Williams D, Bebout J, Zeidman M, Antel J, Munsat T, Roses AD (1990): Chromosome 21 markers in familial amyotrophic lateral sclerosis: *Neurology* 40(Suppl 1):315.

Sobue G, Hashizume Y, Mukai E, Hirayama M, Mitsuma T, Takahashi A (1989):

X-linked recessive bulbospinal neuronopathy: a clinicopathological study. *Brain* 112:209–232.

Thomas FP, Thomas JE, Sadiq SA van den Berg LH, Latov N, Hays AP (1990): Human monoclonal IgM anti-Gal(β1-3)GalNAc autoantibodies bind to the surface of bovine spinal motoneurons. *J Neuropathol Exp Neurol* 49:89–95.

Tierney MC, Fischer RH, Lewis AJ, Zorzitto ML, Snow WG, Reid DW, Niewstraten P (1988): The NINCDS-ADRDA work group criteria for the clinical diagnosis of probable Alzheimer's disease. *Neurology* 38:359–364.

Warner CL, Servidei S, Lange DJ, Miller E, Lovelace RE, Rowland LP (1990): X-linked spinal muscular atrophy (Kennedy's syndrome). A kindred with hypobetalipoproteinemia. *Arch Neurol* 47:1117–1120.

Younger DS, Chou S, Hays AP, Lange DJ, Emerson R, Brin M, Thompson H Jr, Rowland LP (1988): Primary lateral sclerosis. A clinical diagnosis reemerges. *Arch Neurol* 45:1304–1307.

Younger DS, Rowland LP, Latov N, Hays AP, Lange DJ, Sherman W, Inghirami G, Pesce MA, Knowles DM, Powers J, Miller JR, Fetell MR, Lovelace RE (1991): Lymphoma, motor neuron diseases, and amyotrophic lateral sclerosis. *Ann Neurol* 29:78–86.

Ziegler LH (1930): Psychotic and emotional phenomena associated with amyotrophical lateral sclerosis. *Arch Neurol Psychiatry* 24:930–936.